POWER SUPPLY CIRCUITS SOURCEBOOK

Volume 1

Disclaimer

POWER SUPPLY CIRCUITS SOURCEBOOK

Volume 1

 Intellin
Organization

www.intellin.org

Power Supply Circuits Sourcebook Volume 1

Published by the

 Intellin
Organization

www.intellin.org

Copyright 2008 by
Intellin Organization LLC
Publishers - USA

First year of publication 2008
Published in the USA

Disclaimer:

The electronic circuits, software or related documentation in this book are NOT designed nor intended for use (whether free or sold) as on-line control equipment in hazardous environments requiring fail-safe performance, such as, but not limited to, in the operation of nuclear facilities, aircraft navigation or communication systems, air traffic control, direct life support machines or weapons systems in which the failure of the hardware or software could lead directly to death, personal injury, or severe physical or environmental damage ("high risk activities")

The author(s) and publisher(s) take no responsibility for damages or injuries of any kind that may arise from the use or misuse of the electronic circuits in this collection.

The author(s) and publisher(s) specifically disclaim any express or implied warranty or fitness for high risk activities. The electronic circuits, software and related documentation are without warranty of any kind. The author(s) and publisher(s) expressly disclaim all other warranties, express or implied, including, but not limited to, the implied warranties of merchantability and fitness for a particular purpose. Under no circumstances shall the author(s) and publisher(s) be liable for any incidental, special or consequential damages that result from the use or inability to use the circuits and software or related documentation, even if he has been advised of the possibility of such damages.

ISBN 1-4196-9855-9

ISBN13 (EAN13) 9781419698552

Preface

Congratulations for having the first volume of ready-to-apply power supply circuits collection. With this sourcebook, you got the advantage of being able to design and assemble electronic power supply modules fast and worry free. It is a sure way to optimize satisfaction in your hobby. If you are a professional electronic designer, it will help you beat the competition. Speed, efficiency, short development periods, error-free, user and maintenance friendly: these are the factors critical for success. This invaluable book filled with 50 practical ideas will help you beat project deadlines. Make your ideas work!

Make your creativity pay! JUST IN TIME!

informative...
practical...
professional...
versatile...

Acknowledgments

Many Thanks to...

E. Mischa (Optical Recognition)
D. Salinger (Electronics)
P. Schmidt (Cybernetics)
N. Lay (Robotics)
E. Latorilla (Regulation)

Introduction

This sourcebook contains 50 practical electronic circuits for power supply applications. It contains circuits designed as voltage regulators, limiters, current sources, monitors, converters, inverters, and some auxiliary circuits commonly used in power supply devices. You can combine several circuits into one large module to create a powerful electronic device specially designed for your exclusive project. Most circuits are in the voltage regulator category while some are auxiliary that can enhance or protect the power supply circuits.

The transistors used in the circuits have more than one possible equivalents. The pin designations are also shown in details. This feature can help avoid unnecessary delays. The pins shown are either in the bottom view or front view of the transistor unless otherwise noted. Large transistors which cannot or not planned to be installed directly on the PCB must be installed on a heatsink. A dashed circle around a transistor means that the transistor must be heatsinked.

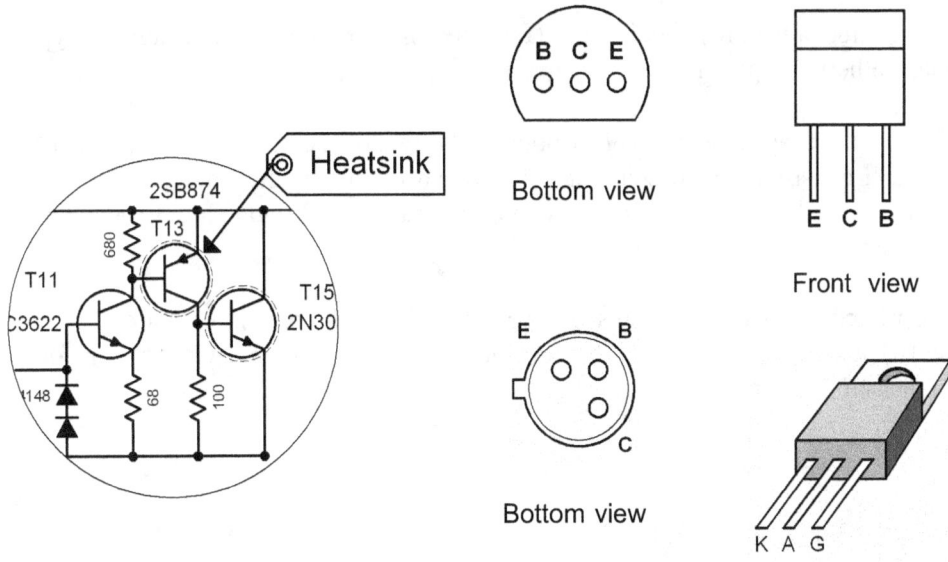

The capacitor values are given in microfarad unless otherwise specified. Electrolytic or polarized capacitors are marked with a plus sign in the diagram. This plus sign corresponds to the capacitor's positive polarity in the circuit. Additionally, their voltage ratings are also given. Nonpolar capacitors are ceramic types and rated with 50 volts.

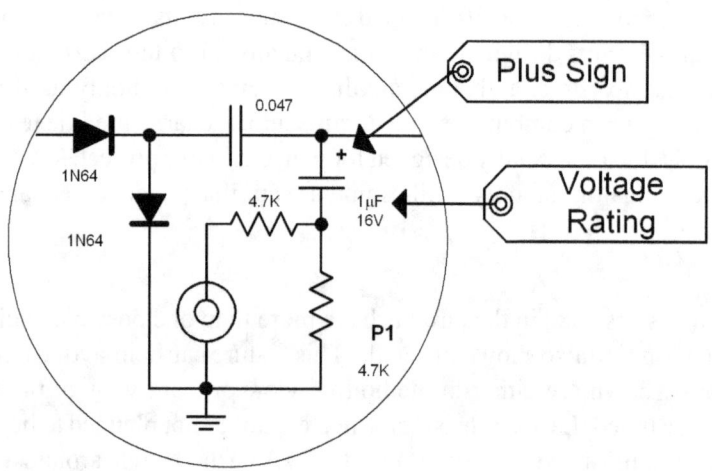

The resistor values are given in ohms (Ω), rated 1/4 watts and are of carbon film type unless otherwise specified.

The appendix pages at the end of the book contain data on semiconductor equivalents including transistors, diodes, zener diodes, FET and other multijunction semiconductors. Two pages of illustrated semiconductor pin designations and layouts are also available as appendix.

The printed circuit board layouts are also printed for the second time at the end of the book. These pages can be cut out for convenience in copying or transferring the layout on the actual pcb board.

Contents

Page number

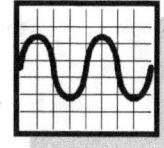

1 HIGH CURRENT REGULATED 5V

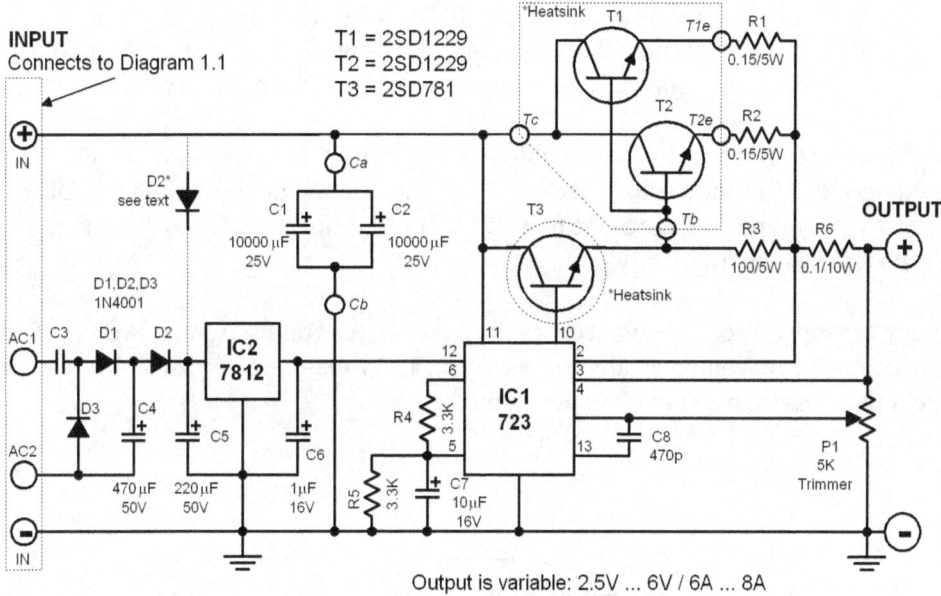

Diagram 1.0   High Current Regulated 5V

   Since the introduction of the 3-pin regulator ICs in the market, the topic of designing power supply circuits has become less and less discussed. However, the same new ICs limit the types of power circuits being built in hobby labs to around 1 ampere units. In addition, power stages are required, because high current (5A or 10A) voltage regulator ICs are very expensive. This is the main reason why hobbyists usually opt for a discrete solution. The idea of cascaded power stages are actually not that bad. For this reason, we have used discrete components for the circuit featured above.

   This regulator circuit was designed for microcomputer systems and other applications that require pulse resistant power supply while delivering high current output at the same time. The 723 IC used in this design is widely replaced by the new 3-pin regulators. However, the 723 IC is adjustable and its technical data are better in some points. It is configured in the circuit to output voltage levels from 2V to 7V. The needed power is produced by doubling the voltage level through a voltage doubler circuit, rectifying it, and regulating it with a 3-pin regulator IC.

This seemingly complicated technique is being applied here to keep the voltage level as low as possible at the secondary winding of the transformer. This makes it possible to keep the power dissipation at transistors T1 and T2 as low as possible too.

The heat sink for T1, T2 and T3 must be properly dimensioned. A heat resistance of around 2°C (CF) is recommended. For the same reason, the resistor values for R1, R2 and R2 must be produced by wiring resistors in parallel. For R1 and R2: connect two 0.33 ohms (5watts) in parallel for each. For R6: connect two 0.22 ohms (5watts) in parallel by a current output of 6 amperes; or connect three 0.22 ohms (5watts) in parallel by an current output of 8 amperes. As additional safety measure, solder these resistors on the pcb with some distance from the board.

If one desires to remove the voltage doubling, just remove the components D1, D3, C3, C4 and move the diode D2 to the positive DC line as shown by the dotted line in the Diagram 1.0. Additionally, the output voltage of the transformer must be increased , and the values of C1, C2, R4 and R5 adjusted appropriately.

This power supply's voltage output drops, by a load of 0.85 ohms, from 5.5 volts to 5.32 volts. That results to a current output of almost 8 amperes! The voltage drop is just 3.3%! The ripple DC voltage is only less than 28 mVeff at the output.

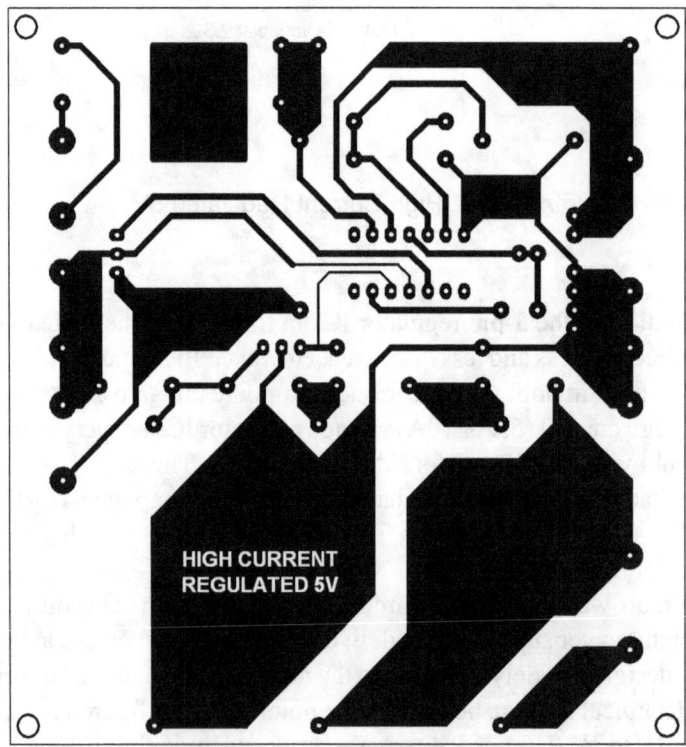

Figure 1.0   Printed Circuit Layout

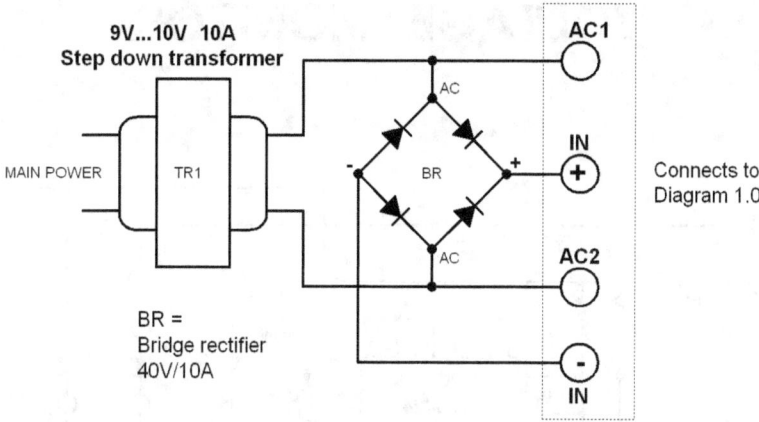

Diagram 1.1   High Current Regulated 5V (AC rectifier part)

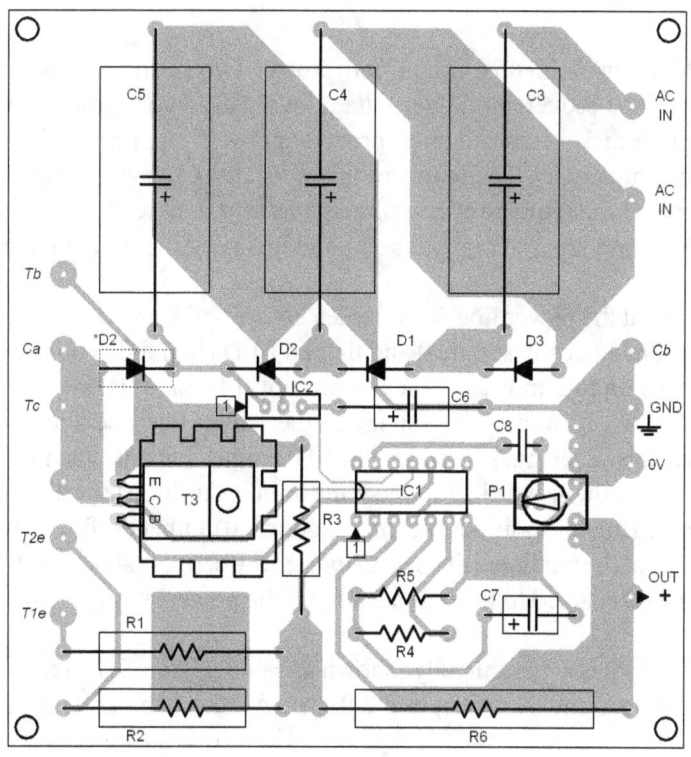

Figure 1.1   Parts Placement Layout

2 VOLTAGE MONITOR

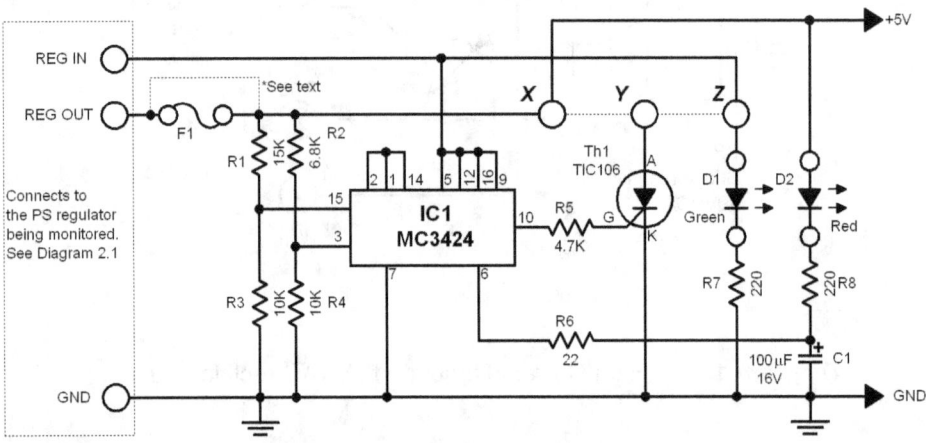

Diagram 2.0 Voltage Monitor

The MC3424 IC used in this circuit can protect two channels from overvoltage levels. Additionally, it can detect over- and undervoltage levels on a 5 volts power line. This makes it very useful as monitor for disk drives and microprocessor power supplies. Each channel of the IC has input and output comparators. The input comparators check the voltage on a 5 volts line. Inside the IC, a constant reference voltage of 2.5 volts is generated. This reference voltage is continuously fed to the non-inverting input of comparator 1 and inverting input of comparator 2.

When the voltage at the power line drops under 2.4 volts, the input comparator of channel one changes its state. Pin 6 becomes logic 0 and the red LED (D2) lights up. D2 can also be used to trigger an interrupt routine in a computer to prompt it to save its data and switch over to an emergency power supply. When the voltage at the power line rise above 6.2 volts, the input comparator of channel 2 changes its state. Pin 10 becomes logic 0. The thyristor TH1 fires and shorts the power line to the ground line. Depending on how the thyristor was connected to the power line (shown as dotted line and points X, Y, Z in Diagram 2.0), either the fuse (F) of the non-regulated line or the fuse (F1) of the regulated line will be busted, effectively shutting off the power line. The diode 1N4001 must be added to the regulator IC of the power supply.

If the power supply has a fuse already (shown as F in Diagram 2.1), replace F1 with a jumper wire and short points Y and Z in Diagram 2.0 with another jumper wire.

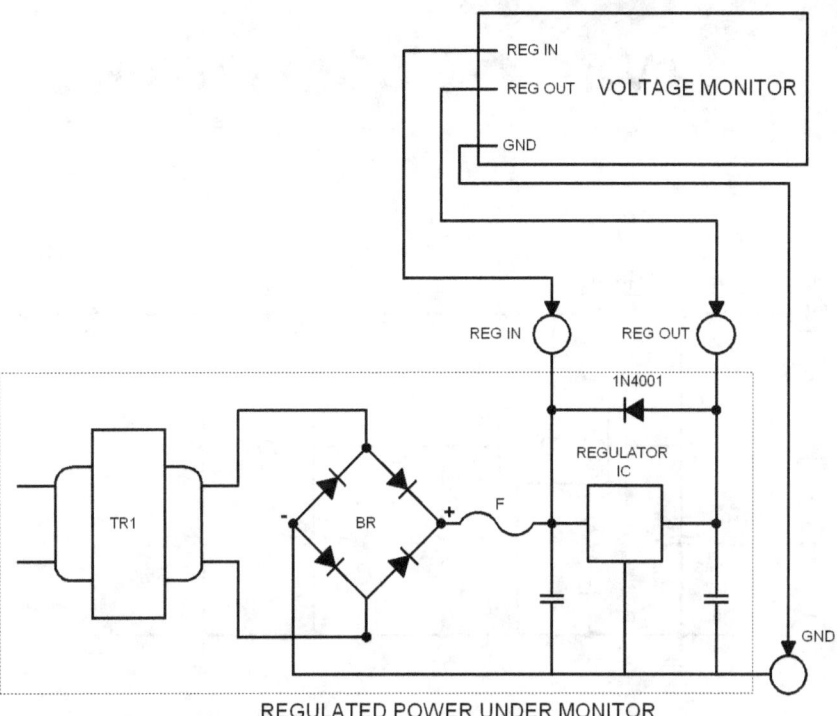

Diagram 2.1   Connecting the voltage monitor to the power supply being monitored

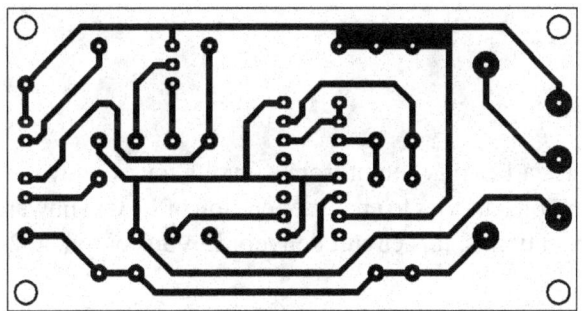

Figure 2.0   Printed Circuit Layout

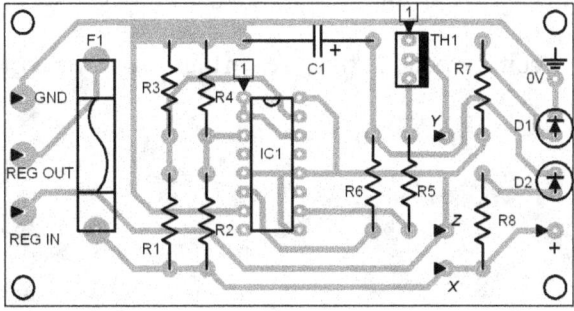

Figure 2.1   Parts Placement Layout

3 6V TO 12V CONVERTER

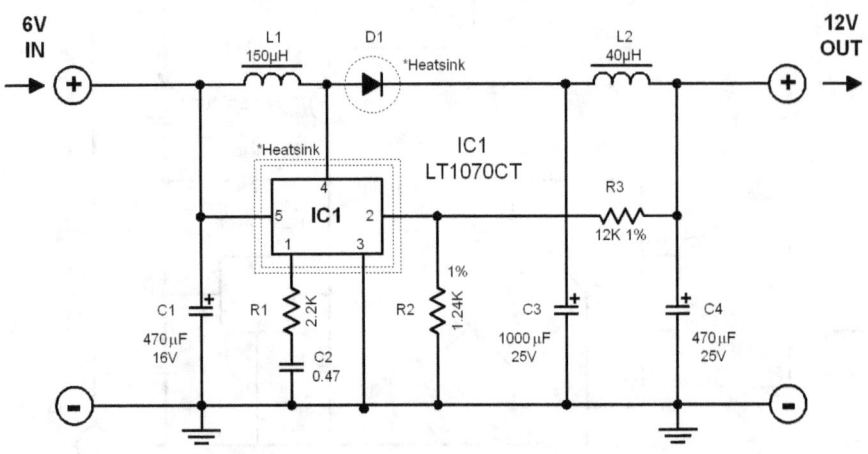

Diagram 3.0   6V to 12V Converter

This circuits generates a 12 volts output from a 6 volts source. It was originally designed to be used in old cars with 6 volt batteries to enable operation of 12 volt powered car radios. No matter what your actual application is, this circuit delivers 12 volts output at 2 amperes maximum.

The current flowing through L1 controls the IC1 that switches on and off at a rate of around 45 kHz. The capacitor C3 gets charged with the counter-inductive voltage from L1 through D1. The LC filter at the output (C4/L2) suppresses any voltage spikes that might appear.

The efficiency of the circuit is around 70%. It is even higher at lower current consumption levels.

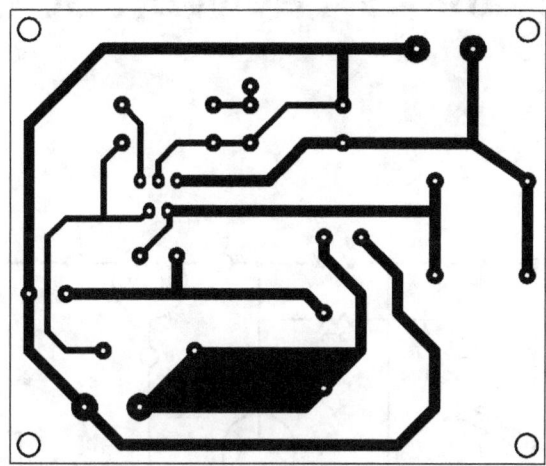

Figure 3.0   Printed Circuit Layout

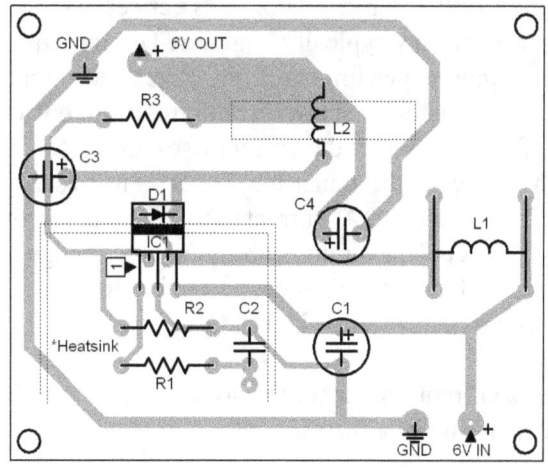

Figure 3.1   Parts Placement Layout

4   5V - 3A POWER SOURCE

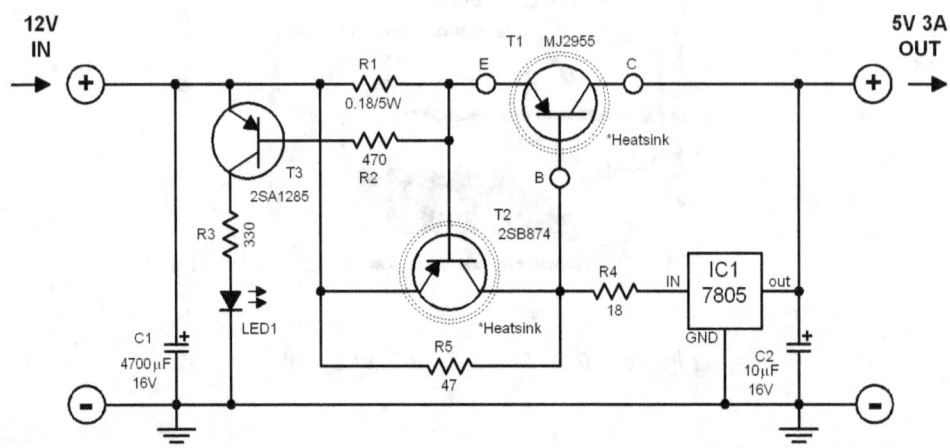

Diagram 4.0   5V - 3A Power Source

For most applications, a 7805 3-pin regulator IC is very easy to use. However, the maximum current output from such unit is low, typically 1 ampere. This low current output is not enough for some applications. To overcome such limitation, the current can be amplified with the addition of a power stage such as the transistor T1 shown in this circuit. At lower current levels, the 7805 IC behaves at its normal function. When the current rises above 15 mA, the voltage drop at R5 becomes high enough to trigger the transistor T1. The transistor T3, on the other, protects transistor T1 in case of a short circuit. At currents above 3 amperes, the voltage drop at R1 is too high that transistor T1 conducts. This limits the base-emitter voltage of T1 and the output current cannot increase any further. Transistor T3 is connected in parallel with transistor T2. The LED1 lights on when the over-current protection kicks in.

Resistor R4 works as a current limiter for the voltage regulator IC1. Without the R4, the IC1 would be heavily loaded in case of a short circuit.

One catch with this circuit: The higher current levels can only flow when the input voltage is higher than 10 volts. In contrast, a single 7805 can function with only 8.5 volts input level. Furthermore, the circuit is not protected from prolonged periods of short circuit.

In constructing the circuit, it is very important to place the transistors T1 and T2 on a properly dimensioned heatsink. A heat resistance of around 3 K/W is recommended. The 7805 IC does not strictly require a heatsink. However, it is a good practice to mount it on a good heatsink too. Certainly, protecting it from high temperature levels would not hurt.

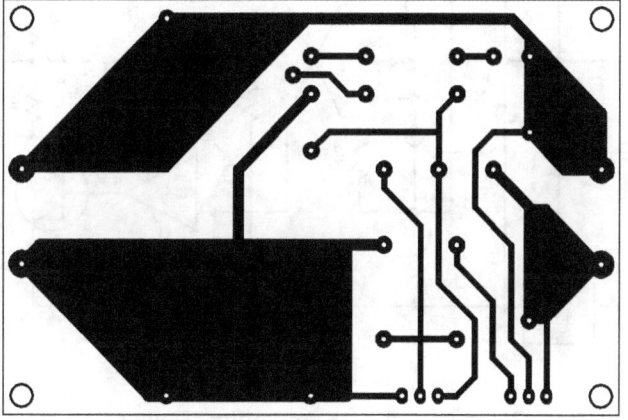

Figure 4.0   Printed Circuit Layout

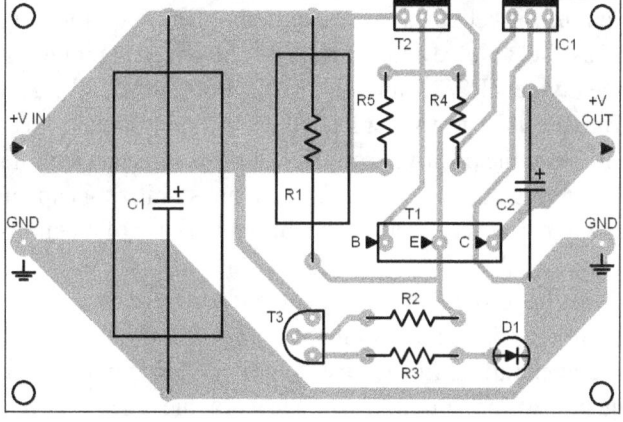

Figure 4.1   Parts Placement Layout

5 SYMMETRICAL POWER SUPPLY

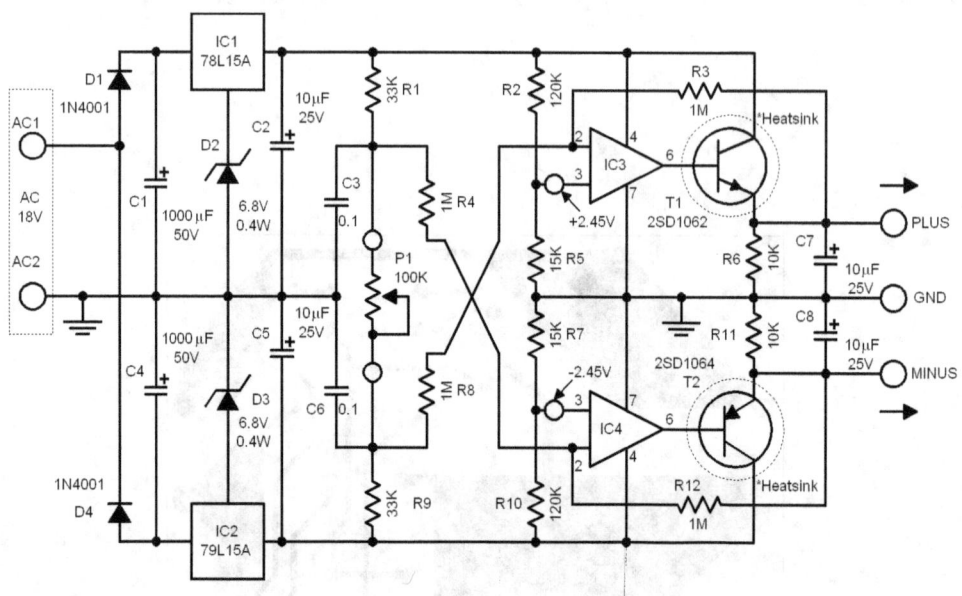

Diagram 5.0   Symmetrical Power Supply

The first part of the circuit on the left side looks typical with its rectifier diodes and filter capacitors C1 and C4. The voltage regulator ICs IC1 and IC2 together with the zener diodes D2 and D3 work to deliver a stable DC voltage of +/- 21.8 volts. Capacitors C2 and C5 stabilize the regulator ICs.

Further into the circuit, it becomes unconventional. Both op-amps together with the driver transistors are wired as DC voltage amplifiers. The non-inverting inputs are connected to +2.45 volts through voltage dividers R2/R5 and -2.45 volts through R7/R10. A regulated symmetrical voltage extracted from the voltage divider P1/R1/R9 and subtracted from the fixed +/- 2.45 volts. The negative from IC3 and the positive from IC4. This results to a symmetrical voltage outputs from +/- 4.9 volts to 18 volts. Capacitors C7 and C8 are stabilizing capacitors. The resistors R6 and R11 are pull-down resistors to preload the outputs in case of an empty load.

The accuracy of the voltage symmetry is dependent on the resistor values of the three voltage divider circuits. If there is a non-symmetry in the output, it is probably due to one of the regulator ICs with its 10% tolerance rating. In such case, adding a 5K trimmer in series with R1 or R9 is recommended. Adjust the trimmer until a symmetry is achieved.

This circuit can deliver a current output up to 100 mA. See page 22 for increasing the current output capacity.

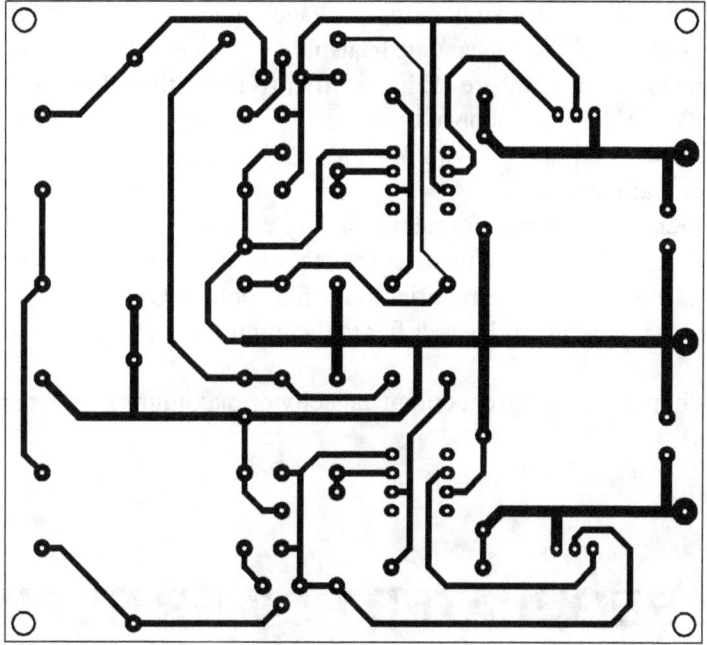

Figure 5.0   Printed Circuit Layout

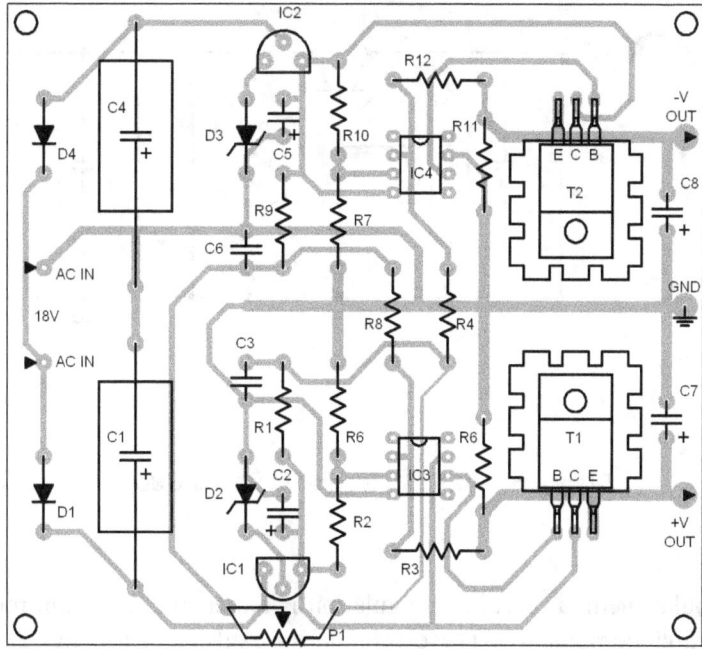

Figure 5.1   Parts Placement Layout

Increasing the current output capacity of the symmetrical power supply:

- Replace the IC1 and IC2 with 1 ampere type voltage regulator ICs (7815 and 7915)
- Mount the new IC1 and IC2 on a separate heatsink
- Replace D1 and D4 with a 1 ampere bridge rectifier (remove them from the pcb)
- Use a 18V-0-18V (1VA) center tapped transformer to supply the AC voltage
- Replace the capacitors C1 and C4 with 4700 μF/45V capacitors
- Use a slow-blow 1 ampere fuse
- Connect the center tap of the transformer secondary with the ground line of the pcb
- Connect the AC pins of the bridge rectifier to both outer taps of the transformer secondary
- Connect the plus pin of the bridge rectifier to the plus pole of C1
- Connect the minus pin of the bridge rectifier to the minus pole of C4

These changes will result to a current output capacity of maximum 1 ampere for the circuit.

6 REGULATED POWER SUPPLY

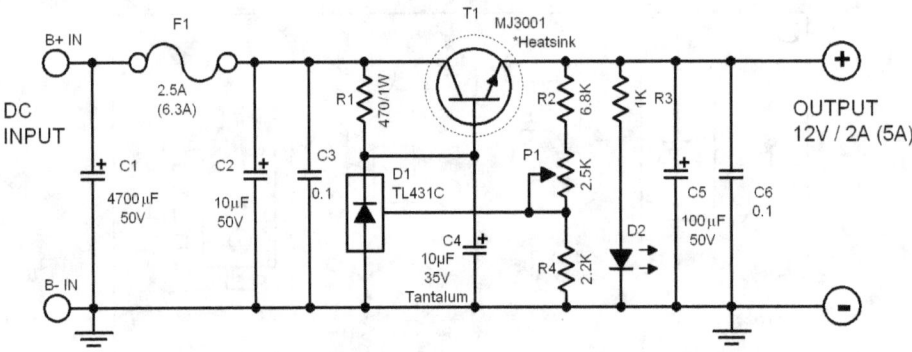

Diagram   6.0   Regulated Power Supply

The most popular method of voltage regulation is either to use a 3-pin regulator IC or the transistor/zener diode combination. These techniques are sufficient in many cases. However, their current output capacity is limited to a maximum of around 1 ampere.

To get higher currents, the regulator must be supported by a power transistor. This, however, affects the regulation characteristics negatively. Ready-made high current regulator ICs (e.g. 5 ampere or 10 ampere types) are very expensive and not easy to find. The transistor-zener diode combination has its limitations too. Its regulation by changing loads and its ripple suppression are not very good.

The above described problems are absent in the featured power supply circuit. Additionally, it is an affordable alternative to expensive high current regulator ICs. At first glance, the circuit looks like the typical transistor/zener diode combination. The decisive difference is in the built-in feedback. This technique reduces the 120 Hz ripple by about 50 dB. This reduction level is not possible in conventional designs.

The reference voltage is supplied by the TL431 (D1 in Diagram 6.0). Diode D1 controls the base of transistor T1 in such a way that the voltage at R4 is a constant 2.5 volts. This makes it easy to calculate the output voltage by using the following formula:

$$U_o = 2.5 \ (1+(p1+R2)/R4) \ \text{in volts}$$

This particular circuit is dimensioned to deliver a 12 volts output. For other output levels, the divider network R2/P1/R4 must be re-dimensioned. Thereby making sure that the current flowing through the divider network does not go below 1 mA.

The power stage is the MJ3001 power transistor. It guarantees a 1000 current gain by 5 ampere collector current. A base current of around 5 mA is needed to control the MJ3001.

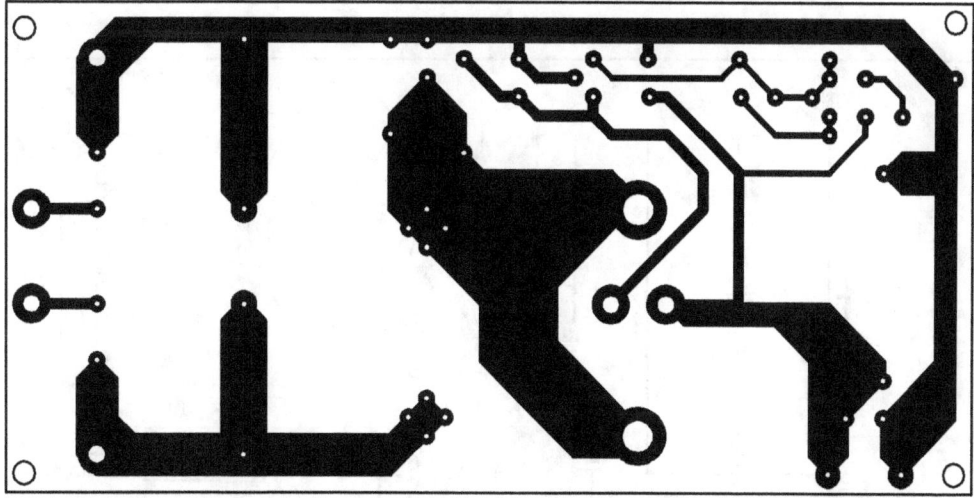

Figure 6.0   Printed Circuit Layout

The current through D1 is at least 0.5 mA. The current through R1 is around 5.5 mA. The value of R1 can be calculated using the formula:

$$R1 = (Uin - Ube - Uo)/IR1$$

where:

Uin is the lowest possible voltage input
Ube is the base-emitter voltage of T1
Uo is the voltage output
IR1 is the R1 current (approx. 5.5 mA)

In practice, a slightly higher value for R1 than the calculated value must be chosen. This is due to the fact that the published current amplification factor of MJ3001 is an average value. This also reduces the power dissipation at D1.

The MJ3001 must be mounted on a heatsink. Select a heatsink that is good enough to handle heat dissipation from the transistor running at least 2 amperes. If one decides to supply a purely DC voltage to the pcb, remove the bridge rectifier and connect the dc supply lines to B+ IN and B- IN respectively. On the other hand, if an AC supply is chosen, use the bridge rectifier and connect the transformer secondary windings to the AC IN connections.

If the circuit is used for 5 ampere current consumptions: Mount the power transistor T1 *externally* on a bigger heatsink. Replace C1 with a 10,000 µF/45V capacitor. If this capacitor does not fit on the pcb, mount it externally. If the consumption stays at 5 amperes on long periods of time or continuously, mount the bridge rectifier externally on a properly sized heatsink. It is also a good idea to use forced air cooling with a mini air blower unit.

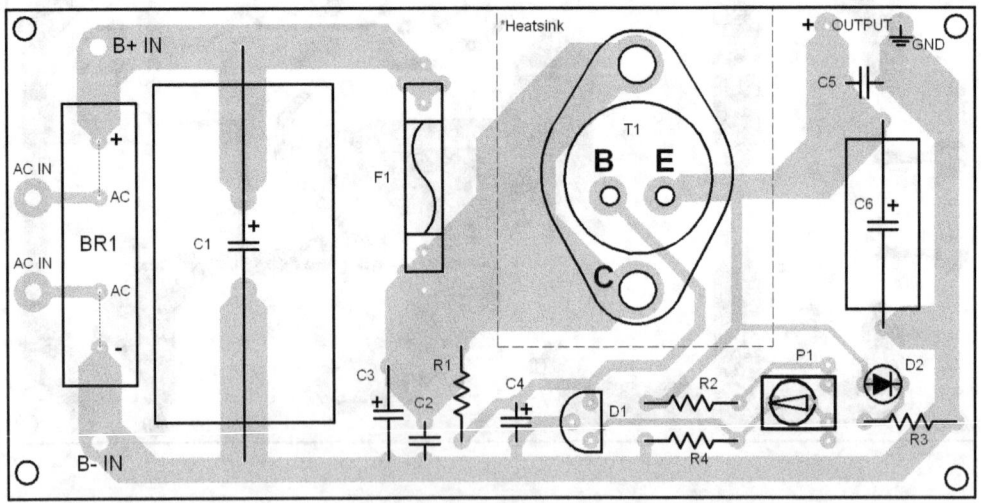

Figure 6.1   Parts Placement Layout for the Regulated Power Supply

7 SINGLE CHIP POWER SUPPLY

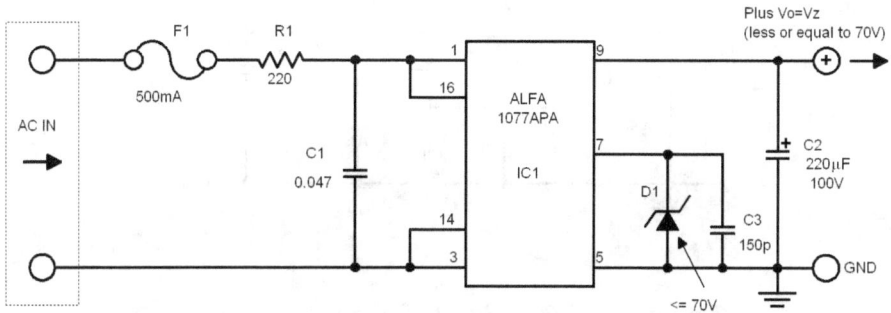

Diagram **7.0** Single Chip Power Supply

The 1077APA IC from Alfa Micro is a fully integrated power supply circuit. It can convert AC voltages from 18Veff up to 276Veff into DC voltages. The chip contains a DC bridge rectifier and a switching amplifier stage. Using this chip, a compact, lightweight and inexpensive power supply can be constructed with a minimum of external components.

The maximum output current is 50 mA, which confines this power supply to low current applications. A single zener diode (D1 in Diagram 7.0) sets the output voltage to a maximum of 70 volts.

The IC goes through a charge and discharge process with each AC cycle. At the start of the cycle, an internal switch connects the capacitor C2 with the internal DC rectified voltage. C2 gets charged to a level set by the zener diode D1. Then the switch opens and stays open until the begin of the next half cycle, while the capacitor C2 delivers current to the load. C2 gets recharged every half cycle. The charging cycle starts again when the DC rectified voltage rises to 1 volt higher than the capacitor voltage of C2. The input frequency can be freely chosen between 48 Hz and 200 Hz. The switching frequency and therefore the charging frequency is always the double of the input frequency.

One obvious disadvantage of the circuit is the total lack of galvanic isolation between the AC mains power supply and the connected load. Encase the entire circuit with a plastic box and use switches with plastic shafts and knobs.

8 SIMPLE SWITCHING SUPPLY

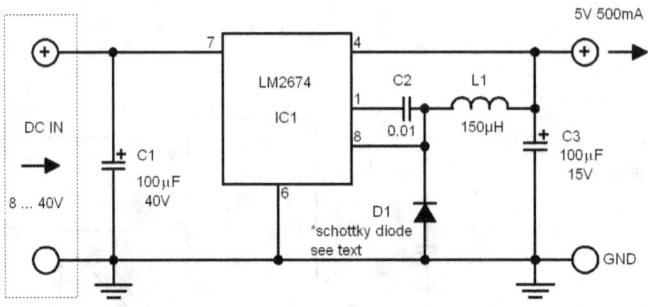

Diagram   8.0   Simple Switching Supply

This simple single chip switching power supply uses an IC from National Semiconductor. This design enables the construction of compact power supplies. The LM2675 is available in different output voltages: 3.3V, 5V and 12V. There is also a version with adjustable output voltage. When applied like the featured circuit above, the chip can deliver a maximum current of 500 mA.

The switching frequency of the IC is 260 kHz. This seemingly high frequency makes it possible to use small coils and capacitors in the application. The efficiency rate is high and the overall dimension of the finished product is small. Under normal circumstances, it is common to achieve an efficiency rate of 90%. In some optimized applications, a 95% efficiency can be achieved.

The chip contains internal protection circuits against very high current outputs and thermal overload. Other positive features of the chip include: soft-start, optional use of an external clock generator to synchronize several voltage converters, and minimize noise.

The featured circuit delivers an output current of 500 mA by a voltage output of 5 volts. Diode D1 is a schottky type with Urev = >45V and maximum current of 3 amperes.

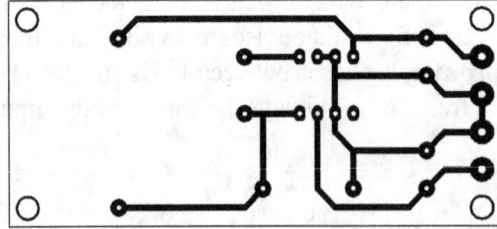

Figure 8.0   Printed Circuit Layout
for the Simple Switching Supply

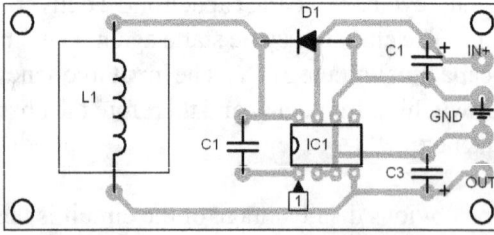

Figure 8.1   Parts Placement Layout
for the Simple Switching Supply

9 COMPACT SYMMETRICAL PS

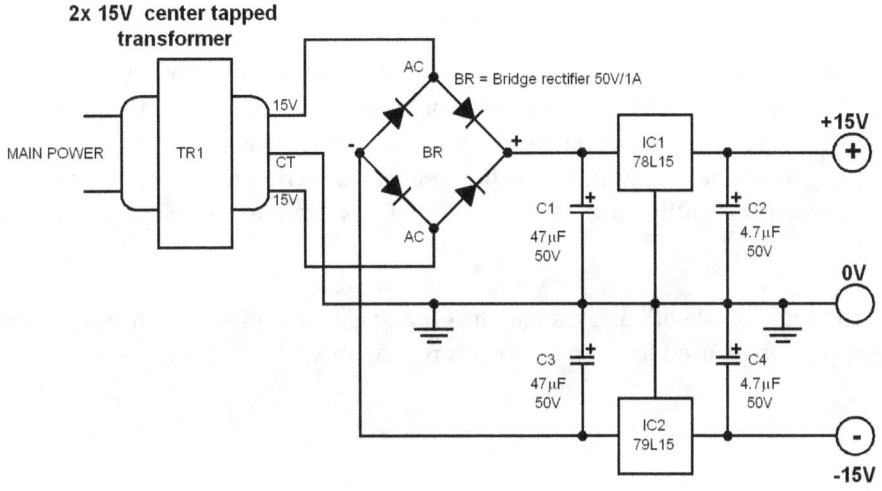

Diagram *9.0* Compact Symmetrical PS

A symmetrical power is usually needed for microphone amplifiers, audio circuits, impedance converters, and most circuits that use opamps. These applications often need only low currents. The circuit featured above offers a symmetrical voltage supply with low current output. It is designed to be compact too.

By an output of +/- 15 volts, it delivers a current of 25 mA. This can also go as high as 100 mA for very short periods of time. By using an transformer with the appropriate secondary voltage, one can use the circuit to produce +/-5V, +/-9V, +/-12V, +/-18V, and +/-24V outputs. The last two voltage levels might prove to be very difficult in finding the correct regulator IC. Due to its compact size, the circuit can be constructed directly inside the application being powered.

One disadvantage of small main transformers is their relatively high internal resistance. This internal resistance results to a big difference between standby output voltage and the voltage under load.

For example, the transformer used in this particular circuit delivers a 32 volts at the input of the regulator IC on standby. Under some circumstances, it is possible that the voltage at the input of the regulator IC is much higher than the allowed level. It is important to note that the upper limits for the regulator ICs are:

  5V = 30V
  12Vand 15V = 35V
  18V and 24V = 40V

One way to avoid a very high standby voltage output is to add resistors in parallel with the transformer's secondary winding. The resistors provide an initial load preventing a very high standby voltage. This resistor, however, must not be low ohmic because otherwise the capacity of the power supply would be very limited, and the resistor would become very hot. In selecting the value of the resistor, a few milliamperes flowing through it is often enough to pull the voltage below the critical level.

If the transformer is rated with 3.5 VA and the C1 and C2 are replaced with 100µF capacitors, the power supply can be used for current loads up to 55 mA.

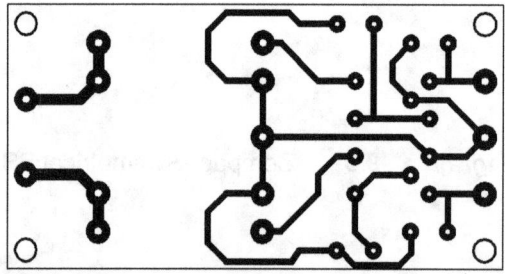

Figure 9.0   Printed Circuit Layout
for the Compact Symmetrical PS

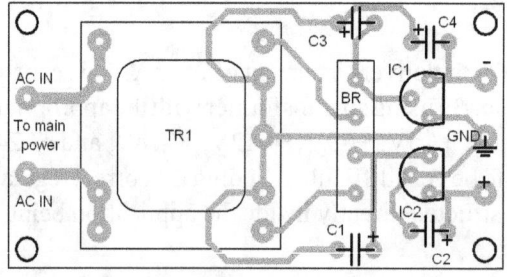

Figure 9.1   Parts Placement Layout
for the Compact Symmetrical PS

10 REMOTE SENSING REGULATOR

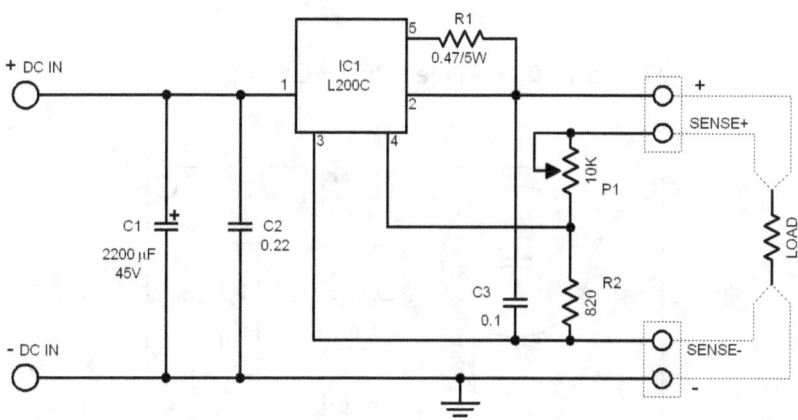

Diagram **10.0** Remote Sensing Regulator

In some applications, it is very important to keep the voltage level constant and highly independent from varying output current levels. These varying current levels can, however, cause voltage regulation problems in many cases. If the load is connected to the regulated power supply with short and high gauge wires, the regulator should not have any problem in keeping the voltage output constant. On the other hand, if the load is connected far away from the power supply with long wires, the voltage drop at the connecting wires could be considerable. This voltage drop will not be sensed by the regulator resulting to a wrong voltage level at the point of the load. Furthermore, this results to an unstable voltage at the load although the voltage at the regulator output is stable.

To solve this problem, the circuit feature above applies a remote sensing technique. The regulator has extra sensing wires connected directly to the load. Since the current flowing through these sensor wires are very low, the voltage drop on these wires are very low too. This way, the regulator circuit can sample the voltage level directly at the load.

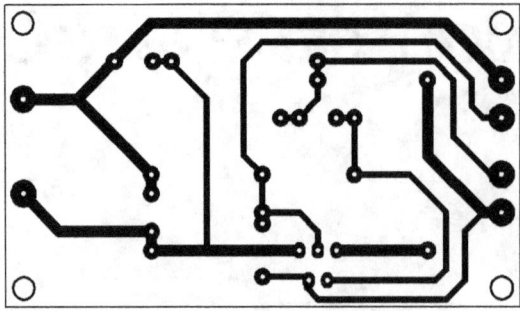

Figure 10.0   Printed Circuit Layout

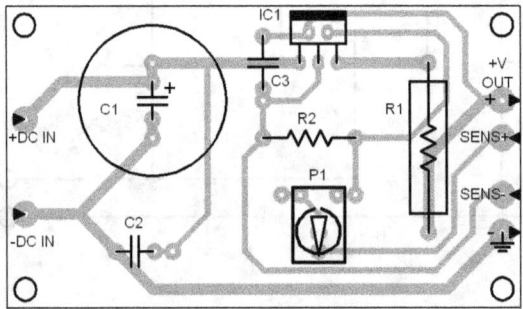

Figure 10.1   Parts Placement Layout

However, most regulated power supplies do not have this remote sensing capability. It is also very difficult to upgrade existing power circuits with such technique. The circuit featured here, on the other hand solves the problem efficiently using the L200 regulator chip.

Output points + and - are the normal power supply outputs. Points SENSE+ and SENSE- are the remote sensing connections. The output voltage can be calculated using this formula:

Uout = 2.77V (I + Rp1/R1)

Using the resistor R2, the maximum current limitation can be calculated with this formula:

Imax = 0.45V/R2

This R2 must be rated to support the high power dissipation levels. The maximum input voltage allowable for the circuit is 40 volts. It delivers a maximum current of 2 amperes.

The IC must be mounted on a properly sized heatsink to protect it from high temperature levels.

11 CURRENT MONITORED SUPPLY

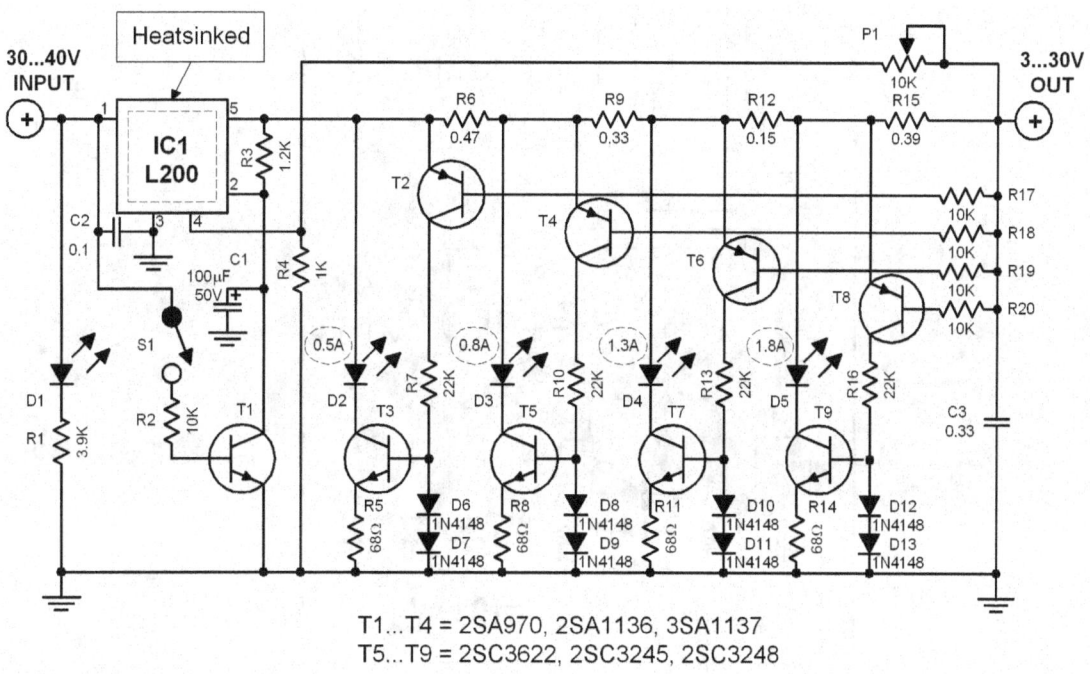

T1...T4 = 2SA970, 2SA1136, 3SA1137
T5...T9 = 2SC3622, 2SC3245, 2SC3248

Diagram 11.0   Current Monitored Power Supply

Here is a reliable power that indicates the actual current being consumed by the load. Voltage regulation is done by the IC L200. Current monitoring is indicated by the LEDs D2...D5 representing current values of 0.5A, 0.8A, 1.3A, and 1.8A. The circuit can supply currents up to 2 A and its output voltage is variable from 3 volts up to 30 volts.

Normally, LED5 is red colored to serve as a warning signal that the maximum current has been reached. The output voltage can be varied through P1. The circuit has a power-on delay composed of R3 and C1. Transistor T1 is the emergency shut-off.

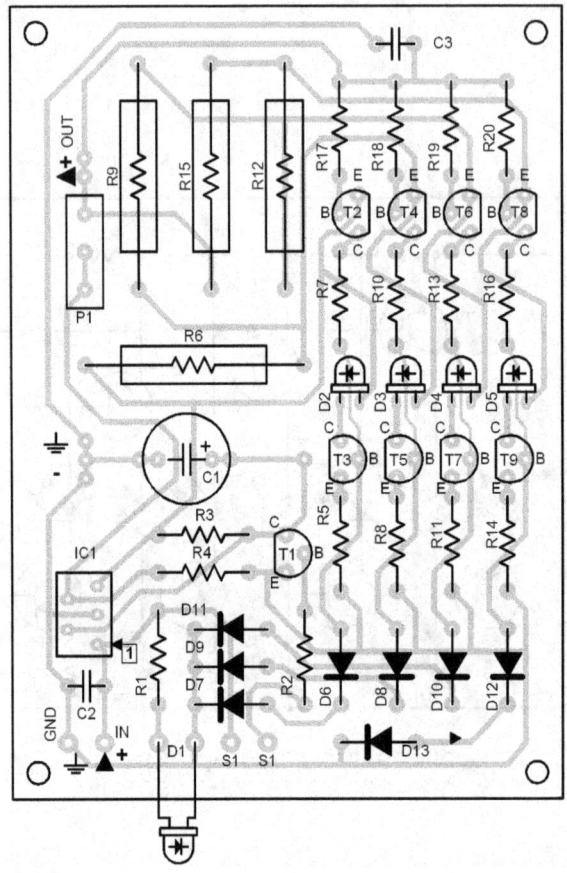

Figure 11.0 Parts Placement Layout

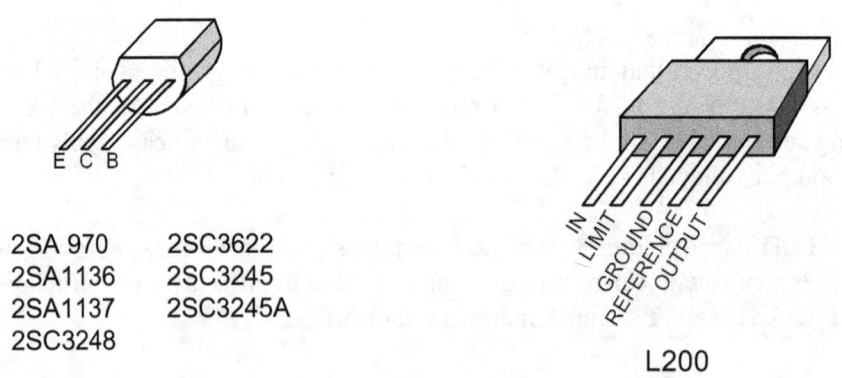

2SA 970	2SC3622
2SA1136	2SC3245
2SA1137	2SC3245A
2SC3248	

L200

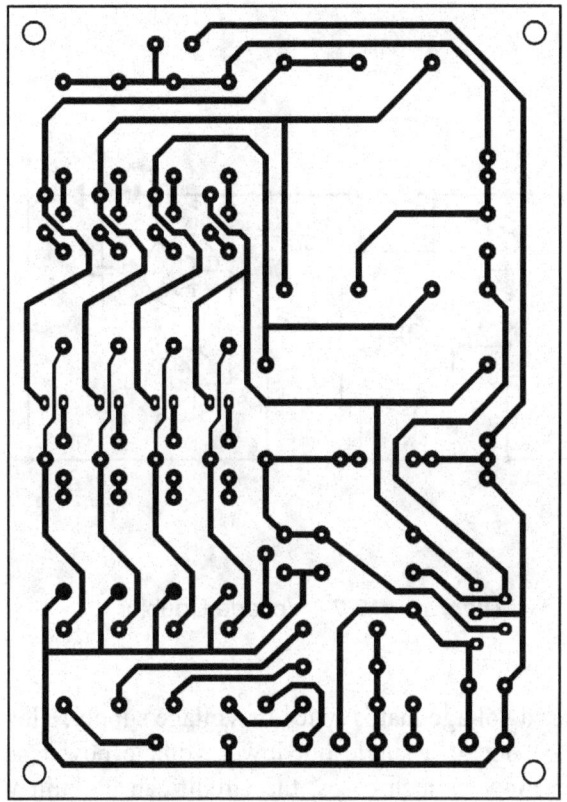

Figure 11.1 Printed Circuit Layout

Parts List:	
R1 = 3.9K	C1 = 100µF/50V
R2,R17,R18,R19,R20 = 10KΩ	C2 = 0.1/50V
R3 = 1.2K	C3 = 0.33/50V
R4 = 1K	D1,D2,D3,D4,D5 = LED
R5,R8,R11,R14 = 68Ω	D6,D7,D8,D9 = 1N4148
R6 = 0.47Ω/5W	D10,D11,D12,D13 = 1N4148
R7,R10,R13,R16 = 22K	T1,T3,T5,T7,T9 = 2SC3622
R9 = 0.33Ω/5W	(2SC3245)(2SC3248)
R12 = 0.15Ω/5W	T2,T4,T6,T8 = 2SA970
R15 = 0.39Ω/5W	(2SA1136)(2SA1137)
	IC1 = L200

12 VOLTAGE DOUBLER

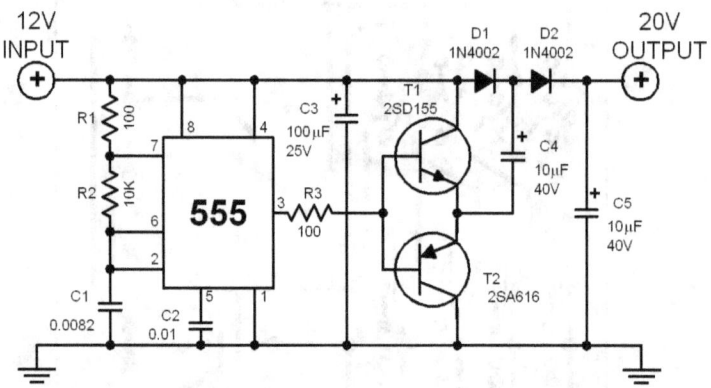

Diagram 12.0   Voltage Doubler

This circuit produces a voltage that is twice its voltage supply. This is useful when a higher voltage level is needed out of a single but lower voltage power supply. Since the current consumption levels are low in such cases, the circuit can be built with minimal number of components.

The electronic circuit is basically a square wave generator using the common 555 timer IC. It is followed by a final stage made of transistors T1 and T2. The actual doubler circuit is made of D1,D2,C4, and C5 components.

The 555 timer IC works as an astable multivibrator and generates a frequency of about 8.5 kHz. The square wave output drives the final stage made of T1 and T2. This is how the doubler works: by a low amplitude of the signal, transistor T1 blocks while T2 conducts. The minus electrode of the capacitor C4 is grounded and charges through D1. By a high amplitude of the signal, transistor T1 conducts while T2 blocks. However, capacitor C4 cannot discharge because it is blocked by D1. The following capacitor C5 is therefore charged with a combined voltage from C4 and the power supply (12V input).

On standby, the circuit delivers around 20 volts. The maximum load must not exceed 70 mA. The actual output voltage is around 18 volts giving an efficiency rating of 32%. On lower current ratings, the voltage is higher. If a stable voltage level is desired, a 3 pin voltage regulator IC can be added at the output. The regulator IC's own current consumption must be added to the total current consumption which must not exceed 70 mA.

13 PS REGULATOR

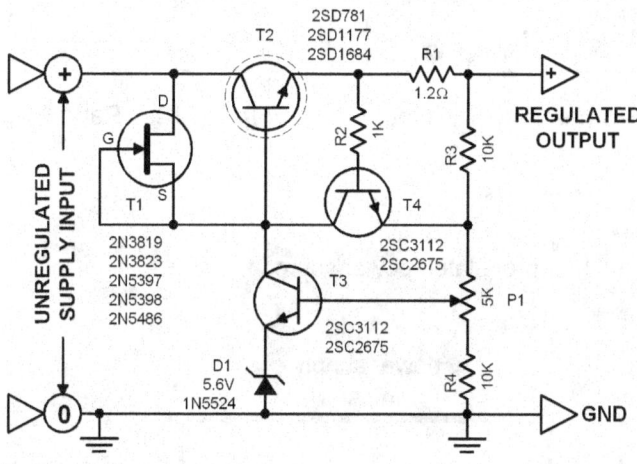

Diagram 13.0   PS Regulator

Most modern power supply regulators are constructed using ICs and work more precisely. However some applications do not really require the precision of an IC. Sometimes discrete components are enough to provide the needed stability. As a hobbyists you are more interested in constructing circuits which present more challenge and chance to understand its working principles. Circuits with discrete components are more fun to work on than ready made, ready to use ICs.

The circuit featured here is designed to give an output voltage of 12V. What is special with this circuit is that it has a current limiter circuit and a constant current source. The current output is limited up to 0.5A. The constant current source (this function is done by a single FET T1) delivers a maximum of 18 mA to the power transistor T1. The output voltage of this power supply is variable through the potentiometer P1. The potentiometer must be wired outside of the circuit board. Be careful in connecting the potentiometer P1 to the circuit board. Never interchange the terminal connections. See Fig. 13.2 for wiring layout.

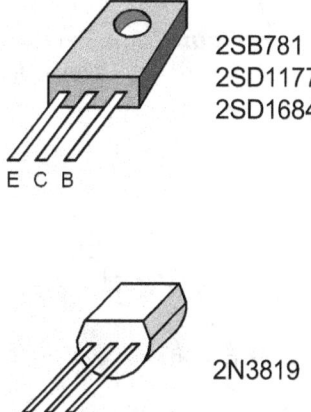

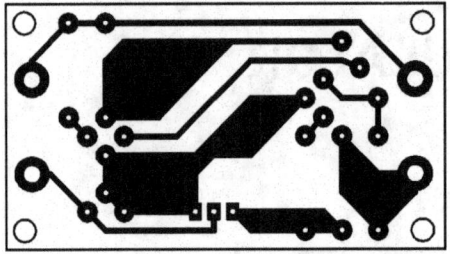

Figure 13.0 Printed Circuit Layout

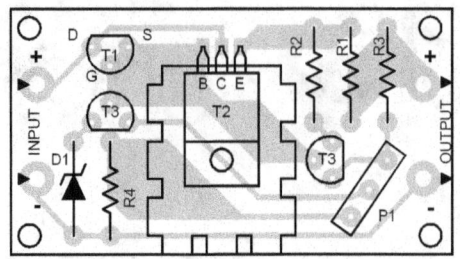

Figure 13.1 Parts Placement Layout

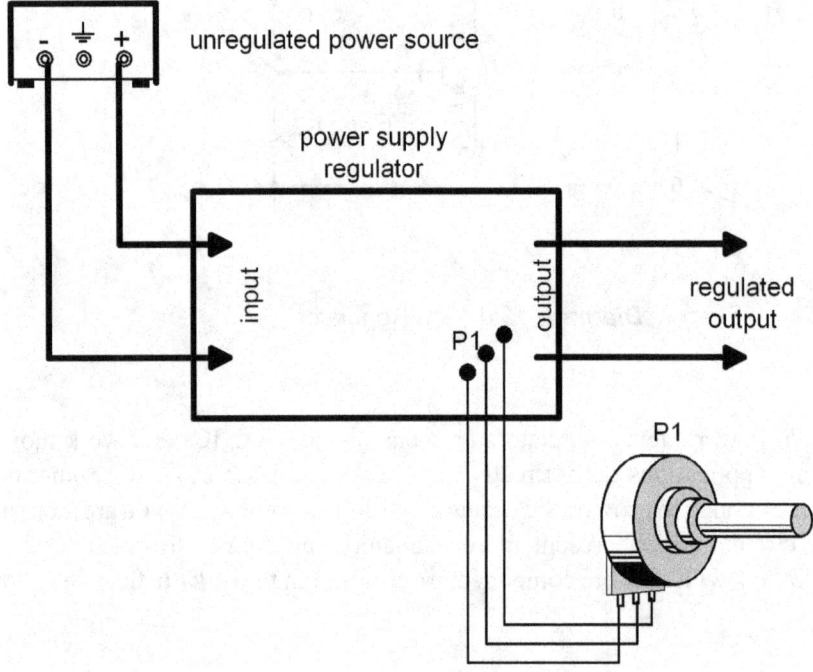

unregulated power source

power supply
regulator

input

output

P1

regulated
output

P1

Figure 13.2 External wiring layout. Take note of the connection of P1.
Never interchange its terminal connections

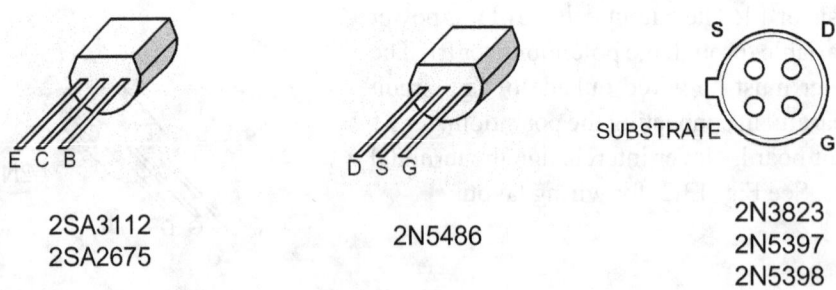

E C B

2SA3112
2SA2675

D S G

2N5486

S   D

SUBSTRATE

G

2N3823
2N5397
2N5398

14   VARIABLE POWER SUPPLY

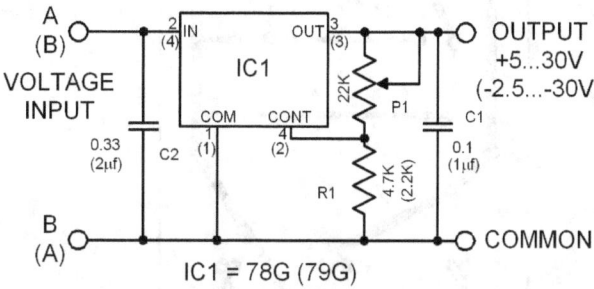

Diagram 14.0   Variable Power Supply

A stable power supply with an adjustable output voltage from 5 volts to 30 volts can be easily constructed with the regulator ICs UA78G or UA79G. These ICs differ from the common three-terminal regulators (which deliver fixed voltage levels only) since their output voltages are adjustable by a voltage level at their control inputs. The maximum current delivered by these ICs is 1 ampere.

The unregulated voltage must be at least 5 volts higher than the desired output level to maintain stability. The input voltage however must not exceed 40 volts. The maximum dissipation of the IC is 15 watts. It has a built-in electrical and thermal overload protection. The circuit featured in the diagram is designed to give a maximum voltage level of 28 volts. If P1 is replaced with 25K potentiometer, the regulator can deliver up to a maximum of 30 volts. Capacitors C1 and C2 stabilize the IC and they must be connected as close as possible to the IC terminals.

To calculate the correct voltage value of the transformer's secondary, use this formula:

where Vts = the voltage output of the transformer's secondary winding and: Vin= is the needed voltage input of the IC.

Formula for Vts
$$Vts= \dfrac{Vin}{0.7}$$

The IC UA79G delivers negative voltage level. Take note that the two ICs have different terminal connections.

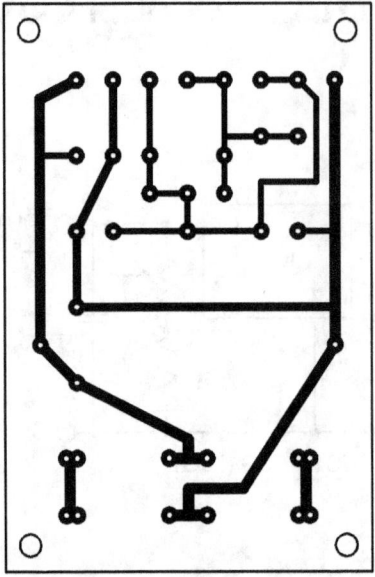

Figure 14.0 Printed Circuit Layout
for the Variable Power Supply

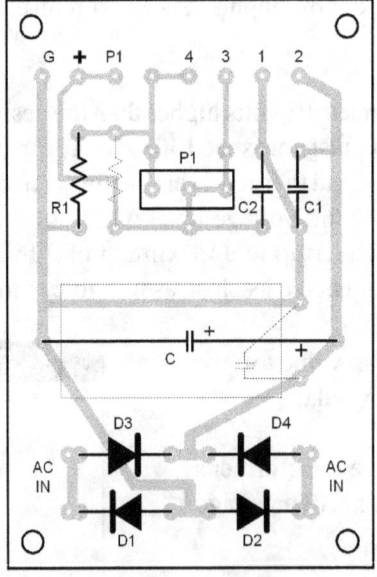

Figure 14.1 Parts Placement
for the Variable Power Supply

15 STABLE POWER SUPPLY

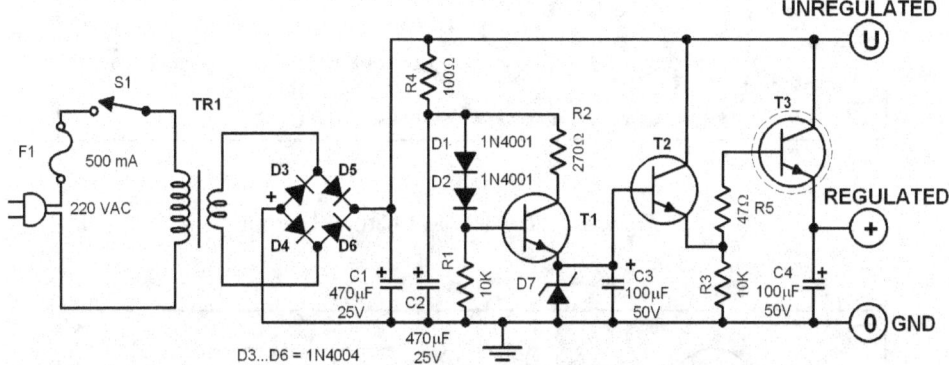

Transistors:
T1 = 2SA1286, 2SA1431, 2SB1288, 2SB1305, 2SB1306
T2 = 2SC3622, 2SC3245, 2SC3248
T3 = 2SD826, 2SD1685, 2SD794

Diagram 15.0   Stable Power Supply

The excellent voltage regulation of this circuit is achieved by feeding a constant current to the zener diode D7. The constant current source is composed of T1,D2,R1,R2. The output voltage can be selected between 3 volts and 47 volts. The output voltage is equal to $Vz+2(0.7V)$. Vz is the zener voltage. The value $2(0.7V)$ comes from the base-emitter voltage of the two transistors T2 and T3. These two transistors work together as emitter follower.

When the circuit is to be used for devices which consume currents higher than 1 ampere, a 2N3055 transistor must be added to it as shown in Diagram 15.1. Transistors T3 and eventually the additional transistor T4 must be properly heatsinked.

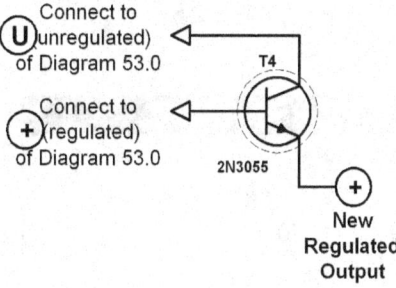

Diagram 15.1   Additional 2N3055

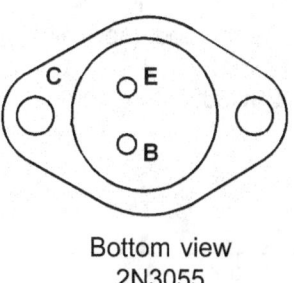

Bottom view
2N3055

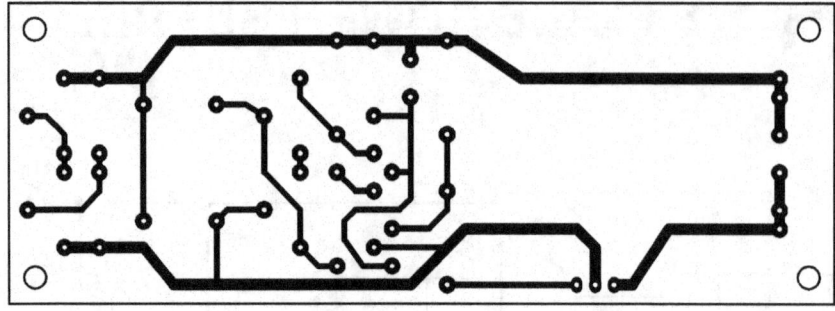

Figure 15.0   Printed Circuit Layout

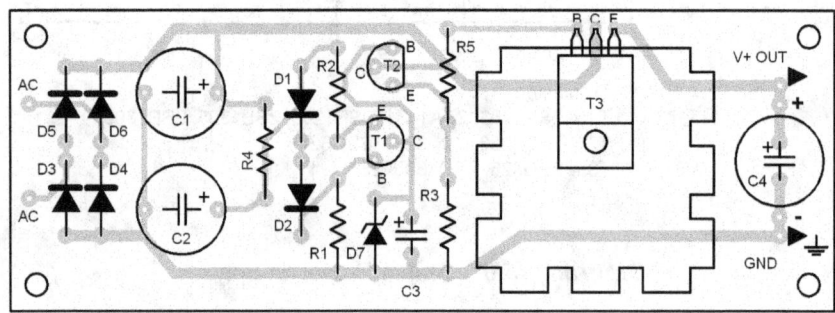

Figure 15.1   Parts Placement Layout

Parts List:

R1,R3 = 10K
R2 = 270Ω
R4 = 100Ω
R5 = 47Ω
C1,C2 = 470µF/25V
C3,C4 = 100µF/25V
D1,D2 = 1N4001
D3,D4,D5,D6= 1N4004
D7 = Zener diode(see text)
T1 = 2SA1286
T2 = 2SC3622
T3 = 2SD826

E C B

2SC3622   2SA1286
2SC3245   2SB1288
2SC3245A   2SB1305
2SC3248   2SB1306

E C B

2SD826
2SD794

2SA143

E C B

2SD1682
2SD1685

16 STABLE ZENER VOLTAGE

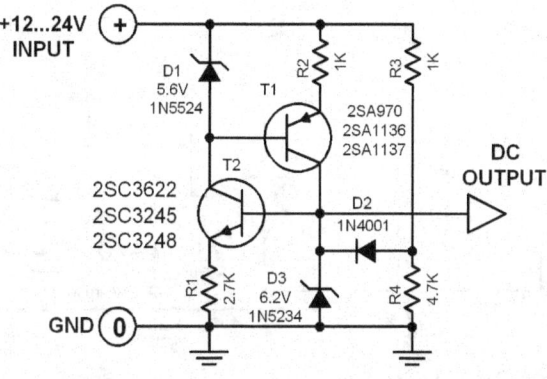

Diagram 16.0 Stable Z-Voltage Source

The zener voltage of a zener diode depends on its bias current and can vary from type to type and from capacity to capacity. It presents a problem for circuits which need a highly stable reference voltage. The solution is to make the bias current constant so that the zener voltage also remains constant.

The circuit featured here does just that. It uses a 6.2 volt zener diode. Zener diodes with different zener voltages can also be used as long as the values of R1... R4 are changed accordingly.

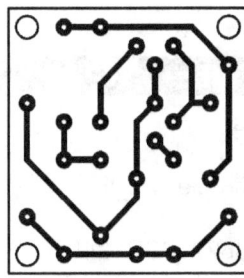

Figure 16.0
Printed Circuit Layout

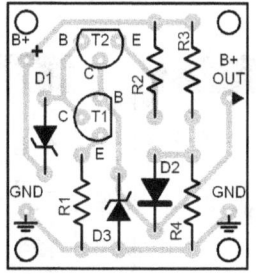

Figure 16.1
Parts Placement Layout

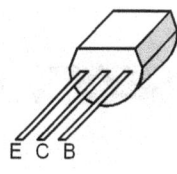

2SA970
2SA1136
2SA1137
2SC3622
2SC3245
2SC3248

17 3-WATT AMP POWER SUPPLY

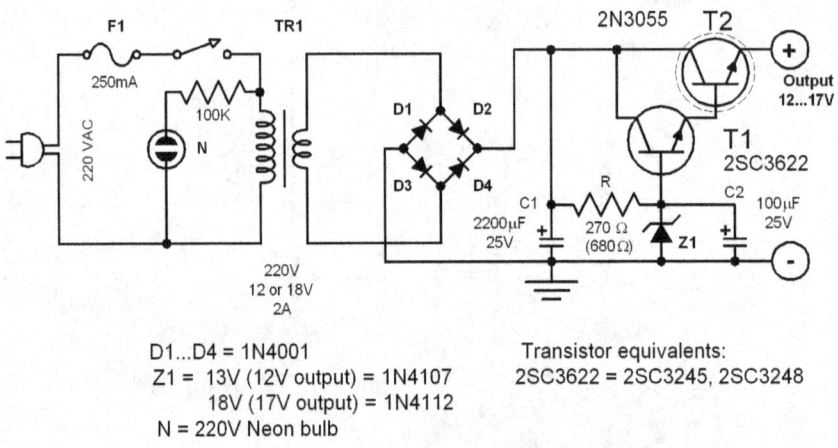

Diagram 17.0 3-Watt Amplifier Power Supply

D1...D4 = 1N4001
Z1 = 13V (12V output) = 1N4107
18V (17V output) = 1N4112
N = 220V Neon bulb

Transistor equivalents:
2SC3622 = 2SC3245, 2SC3248

The power stage of this circuit is composed of two transistors T1 and T2 in darlington configuration. The regulation is controlled by the zener diode Z1. The voltage of the zener diode must be 1 volt higher than the desired output voltage. Table 17.0 shows the component values for two voltage levels.

Table 17.0		
	12V	**17V**
R1	270W	680W
Z1	13V	18V
Tr1	12V	18V
T2	2SC3622	2SC3245A

Parts List:	
R1=	270 ohms
C1=	2,200µF/25V electrolytic
C2=	100µF/25v electrolytic
T1=	2N3055
T2=	2SC2SC3622 (2SC3245)
Tr=	220V/12V, 2A
D1...D4= 2A/100V diode or bridge rectifier with similar specifications.	

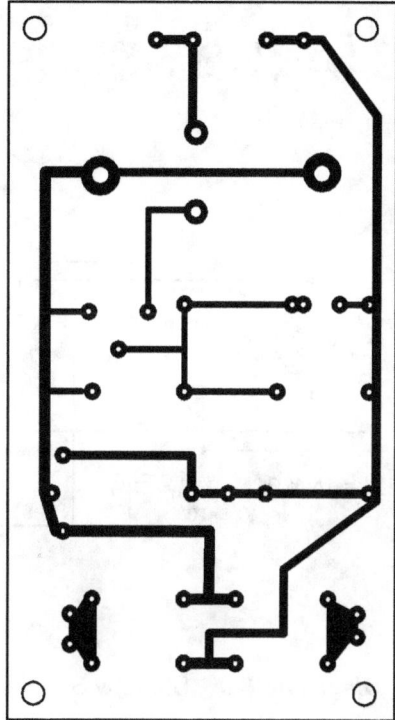

Figure 17.0 Printed Circuit Layout

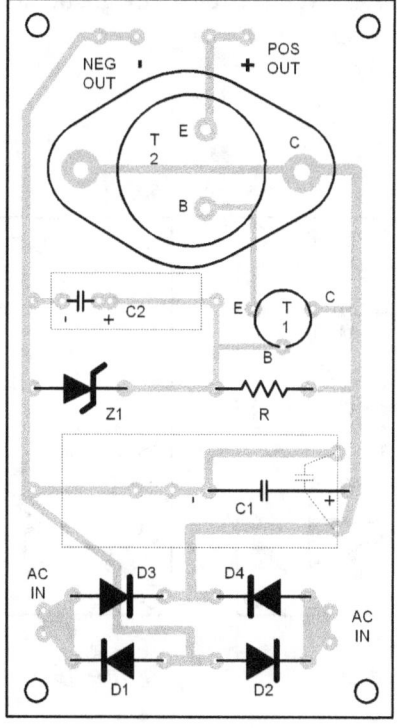

Figure 17.1 Parts Placement Layout

18   STANDBY SUPPLY

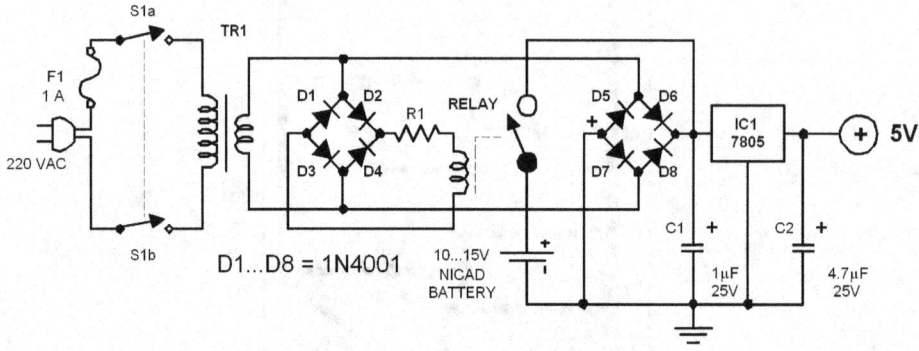

Diagram 18.0   Standby Supply

This circuit provides continuous power supply for memory circuits during sudden power break downs. When the main line loses power, the relay's magnetic field collapses. Since the contact is a normally-closed (NC) type, it connects the standby battery cells to the regulator circuit thereby providing continuous power.

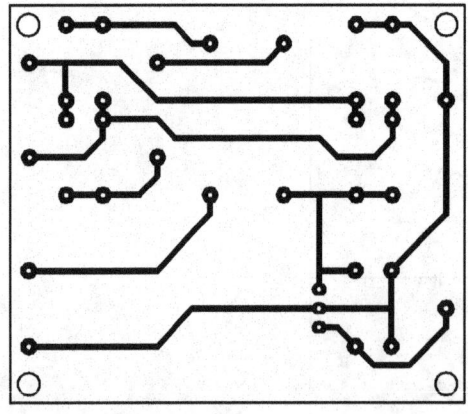

Figure 18.0   Printed Circuit Layout

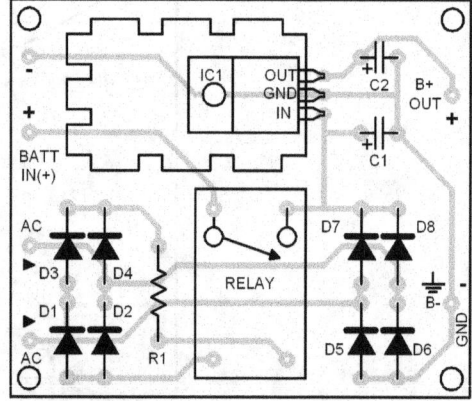

Figure 18.1   Parts Placement Layout

19 DC REGULATOR

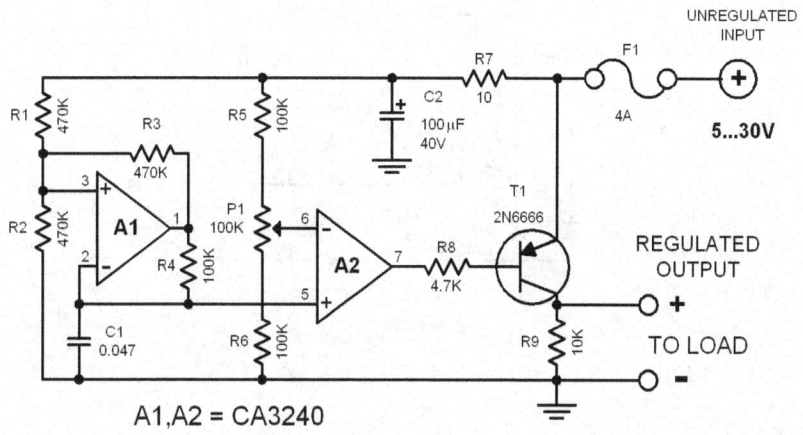

Diagram 19.0 DC Regulator

This circuit regulates a DC power output. It has a very wide application range. It can be used to control the speed of a motor, a pump, a toy train, the brightness of a LED or lamp, etc. Practically, it can be used in any application that uses a regulated DC power with pulse width modulation (PWM). It was originally used as a jumbo LED dimmer.

The circuit works this way: the A1 opamp functions as a square wave generator. At its non-inverting input is a by product triangle wave signal (almost triangle anyway). The IC A2 following it functions as a simple comparator. A reference voltage is fed to the inverting input of the A2 IC via the potentiometer P1. The output of A2 is a square wave signal with a constant frequency of around 200 Hertz. This signal has a variable pulse width from 0 to 100%. The P1 sets the trigger point of the pulse. The transistor T1 works as the actual regulator by switching a relatively high current with a maximum of 5 amperes. The power supply voltage must be between 5 volts (minimum) and 30 volts (maximum). Take note that the lower the supply voltage is, the lower the efficiency of the circuit becomes.

20 SYMMETRICAL AUX PS

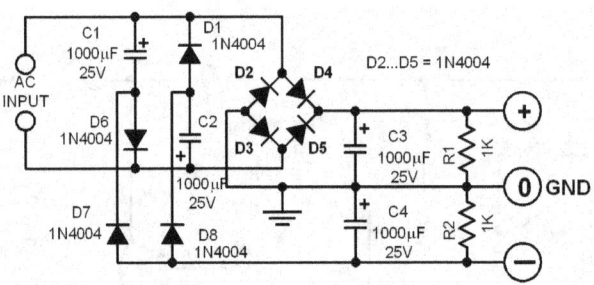

Diagram 20.0 Symmetrical Aux PS

This simple circuit provides a symmetrical power supply derived from the AC output of a transformer's secondary winding without center tap. It is very useful in applications where both positive and negative lines are needed but a center tapped transformer is not available.

The circuit is basically made of a bridge rectifier with additional components that function as the negative line. Take note however that this negative line can deliver less current than the positive line.

The resistors R1 and R2 are initial loads. They are added to make sure that the negative line is active even if the circuit has no load attached to its output lines. This trick is applied here because, otherwise, no negative output will be produced when the positive line has no load. In case you need the negative line to deliver more current than the positive line, just reverse the circuit.

To do this, just reverse the polarities of all diodes and electrolytic capacitors. It means all diodes and electrolytic caps, no exemption. The bridge rectifier will deliver more power in the negative line in such case.

The actual circuit can handle AC inputs up to a maximum of 18 volts AC. If you need a higher voltage level, change the voltage ratings of the diodes and capacitors to at least the double of the maximum AC input voltage. To select the correct diodes, see the appropriate table at the end pages of this book.

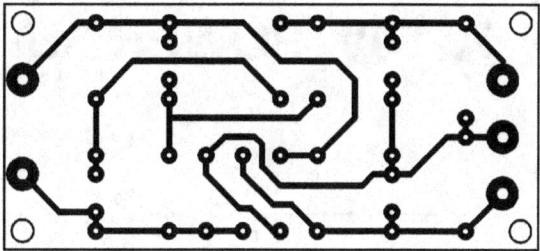

Figure 20.0 Printed Circuit Layout

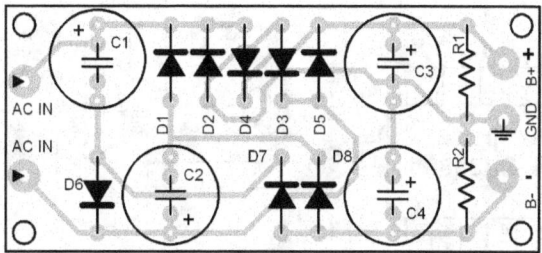

Figure 20.1 Parts Placement Layout

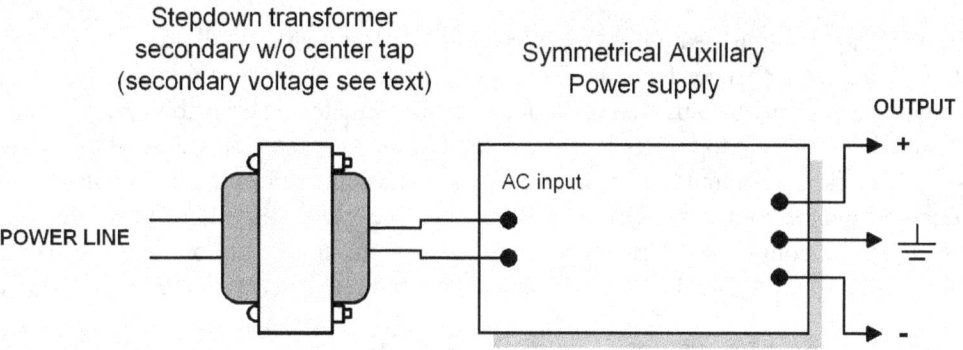

Figure 20.2 External wiring layout for the symmetrical power supply. Take note that the negative line delivers less current than the positive line.

21   0.1V-50V POWER SUPPLY

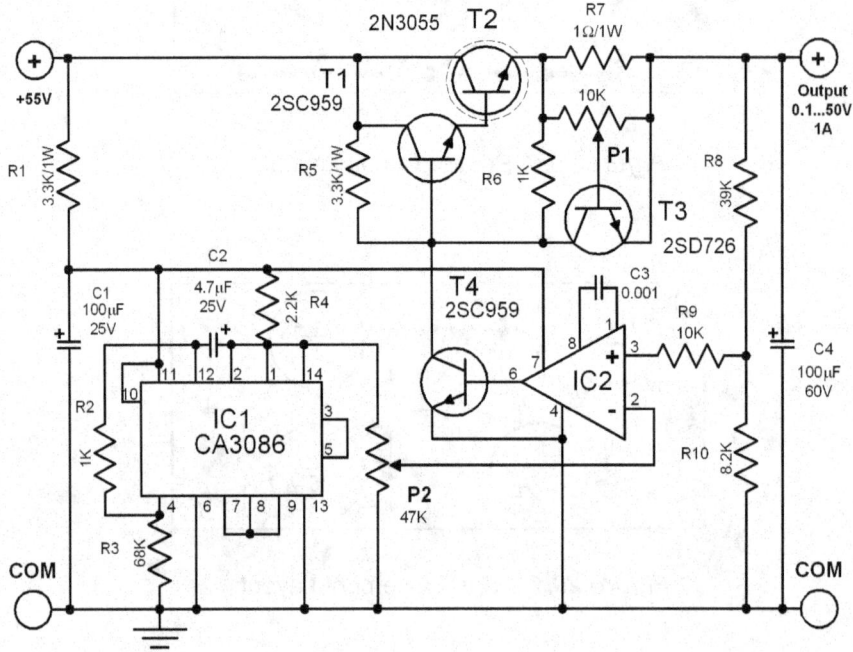

Diagram 21.0   0.1V-50V Power Supply

This power supply circuit is highly stabilized that its output voltage will drop by only 0.005% even though the load changes from 0 to 100 %. Another excellent capability is that the output voltage will change only by 0.01 % if the input voltage fluctuates. The capability of the circuit to be adjusted from 0.1V up to 50V is due to the application of opamp IC CA3130 in the circuit. Transistor T4 raises the output voltage to higher level, and at the same time it separates the lower level opamp from the high level of the output voltage. The reference voltage is supplied by IC1. It is a temperature compensated transistor array with 5 transistors. Four of these transistors are used as reference diodes, and the fifth one sets the output impedance of the reference source.

The reference voltage is set through P1. The opamp CA3130 compares the reference voltage at its minus input to the output voltage at its plus input. The output voltage passes first through a voltage divider before it is fed into the plus input of the opamp. Transistors T1 and T2 work as darlington pair and amplifies the current. Transistor T3 functions as current limiter. The current limit is adjustable through P1, and the lowest current limit is 0.6 ampere. Once potentiometer P1 is set at maximum, current limiting is disabled.

22 DC TO DC CONVERTER

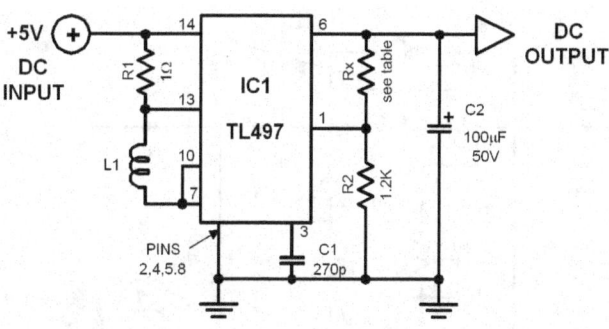

Diagram 22.0 DC to DC Converter

In some applications, one needs a voltage level that is higher than the one immediately available, like in programming EPROMS: a 25V is necessary, but common digital circuits have only 5V and 12V lines. This converter circuit solves that problem. It converts a DC voltage to a much higher level. To be able to obtain the desired output level, use Table 22.1 to know the correct value of Rx.

L1 is a small coil with 85 turns of 0.2mm magnet wire around a ferrite core. Its total inductance must be around 100 µH.

DC IN	DC OUT	I max	R x
5	10	125 mA	8.8
5	15	80 mA	13.8
5	20	60 mA	18.8
5	25	50 mA	23.8

Table 22.1 Value of Rx

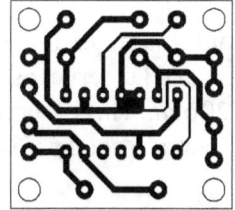

Figure 22.0
Printed Circuit Layout

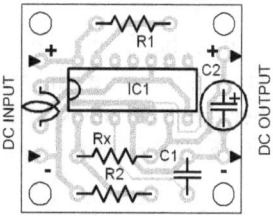

Figure 22.1
Parts Placement Layout

23 3-AMPERE POWER SUPPLY

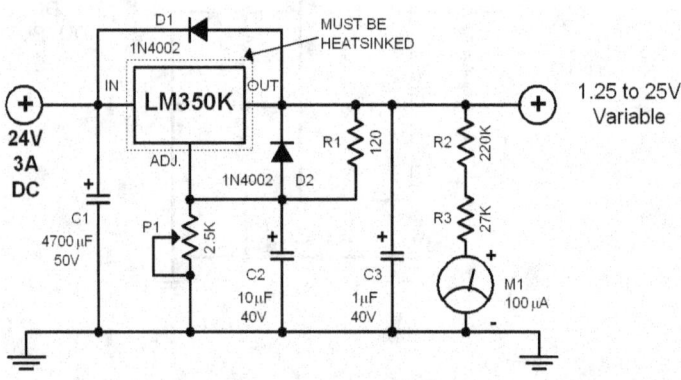

Diagram 23.0 3-Ampere Power Supply

This is a compact power supply that delivers a stable voltage and currents up to 3 amperes. The circuit is very conventional and its voltage output can be varied from 1.25 volts to 25 volts.

The main module is the LM350 IC which integrates a voltage regulator and a power stage. It also has a built-in overload protection which activates at 30 watts of power dissipation. The voltage output is set by connecting the „adj" pin of the IC to the voltage divider made of R1 and P1. The output voltage can be calculated using the following formula:

1.25 V x (1 + P1/R1)

where the P1 value is between 0 and 2.5 kiloohm. Capacitor C1 is a common ripple filter while capacitors C2 and C3 improves the regulation. The diodes D1 and D2 serves as protection for the regulator IC when the IC output is turned off. Resistor R1 is 120 ohm. This ensures that the minimal load current for the IC (around 3.4 mA) is high enough to maintain good performance.

One thing that is most important in building the electronic circuit: provide adequate heatsink for the LM350 IC. The power dissipation at the IC is very high, around 85 watts.

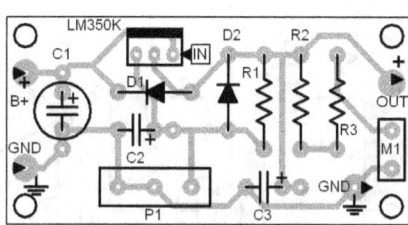

Figure 23.0 Printed Circuit Layout

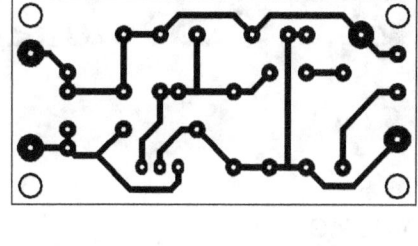

LM350K

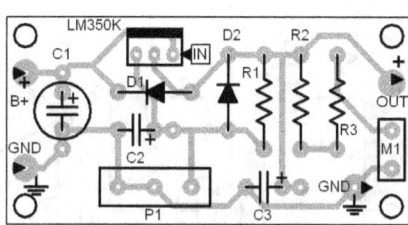

Figure 23.1 Parts Placement Layout

Consider that the heat resistance of a TO-3 package is 1.5°C/W and the maximum allowable temperature is 150°C. If a heatsink is used with a heat resistance of 1.5°C/W, then the total heat resistance is 4°C/W. At 30 watts dissipation and 25°C outside temperature, the resulting internal IC temperature is 145°C. Once this dissipation level is reached, the internal protection activates shutting down the IC.

One way to avoid high dissipation levels is to use a lower voltage transformers when the needed output voltages are low. To put it simply: if you are using the circuit to supply voltages around 9 volts, do not use a 25 volts transformer but use a lower voltage one instead (e.g. 12 or 15 volts).

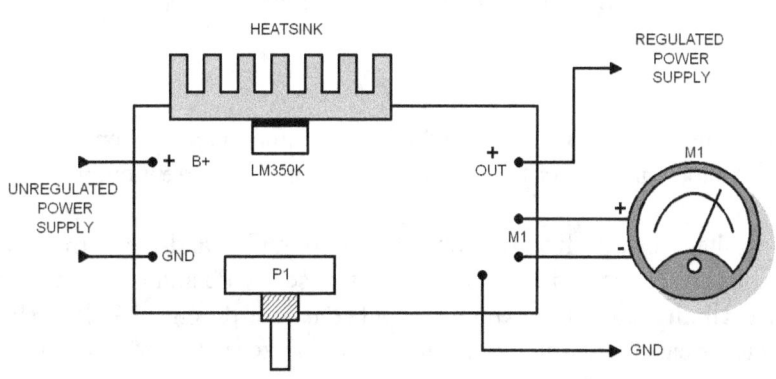

Figure 23.2 External Wiring and Heatsink Layout

24 0..50V/0..2A POWER SUPPLY

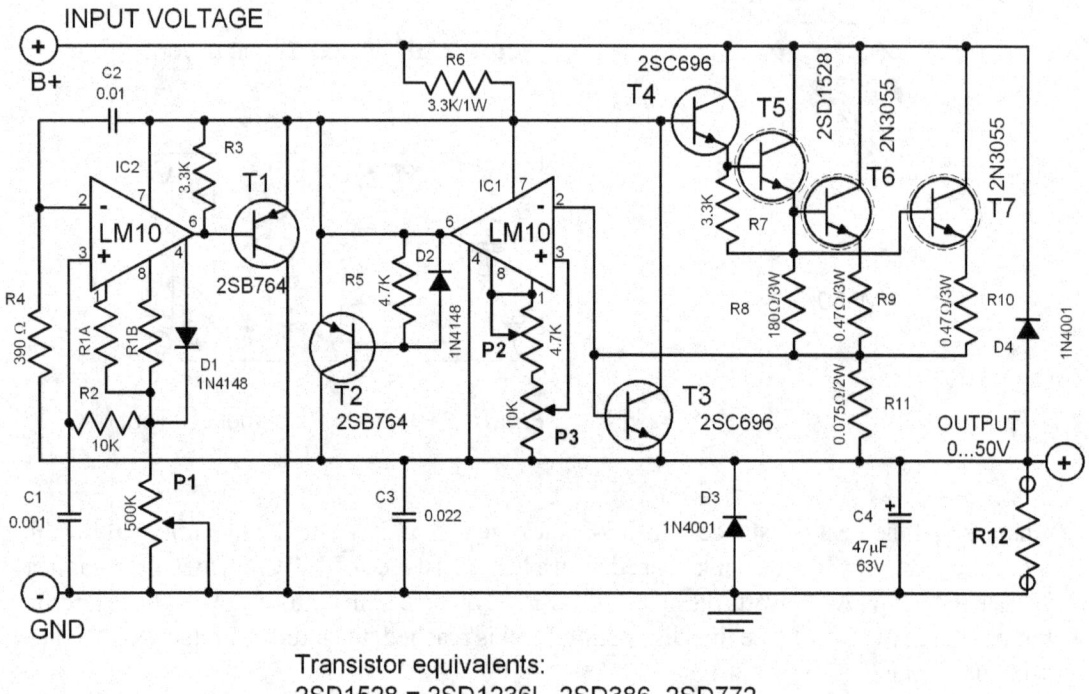

Transistor equivalents:
2SD1528 = 2SD1236L, 2SD386, 2SD772
2SC696 = 2SC696A, 2SD1639, 2N1990
2SB764 = 2SA1705, 2SB1116, 2SA1706

Diagram 24.0   0..50V/0..2A Power Supply

   This power supply uses two LM10 ICs which has an integrated reference source. The circuit is designed to be short-circuit proof with variable voltage and current outputs.

   The output voltage can be linearly controlled through P1 and the current can be linearly controlled through P3. The maximum current limit is set by P2 and is adjusted only once. The maximum current limit is 2A. The maximum output voltage on the other hand is set by a fixed-value resistor connected in parallel to potentiometer P1. Using a fixed value resistor for this purpose helps in reducing noise at the output voltage.

The voltage regulation works this way:

The minus input of IC1 is connected to the output voltage and the plus input is connected to the combination R1/P1. This IC controls the base of transistor T1 so that any voltage difference between the two inputs are equalized. The collector current of this transistor causes a voltage drop at R6 which controls the output voltage through the darlington stage. Pin 1 of LM10 chip is the reference output. Once the output has stabilized to the desired level, the difference between the two inputs is 0.

The voltage level at the junction of R1/P1 is then equal to the voltage level at the minus input of this opamp. Once the reference level is changed by adjusting the potentiometer P1, a voltage difference between the two inputs is sensed by the opamp. The opamp reacts by changing the control voltage until the difference voltage is pulled back to 0. The current is controlled by another LM10 chip. To stabilize the current, a reference level taken from potentiometer P3 is compared with the voltage at resistor R11. The current flows through this resistor. Since LM10 is not a fast opamp, the current is limited through T3. T3 limits the current to the maximum of 2 amperes.

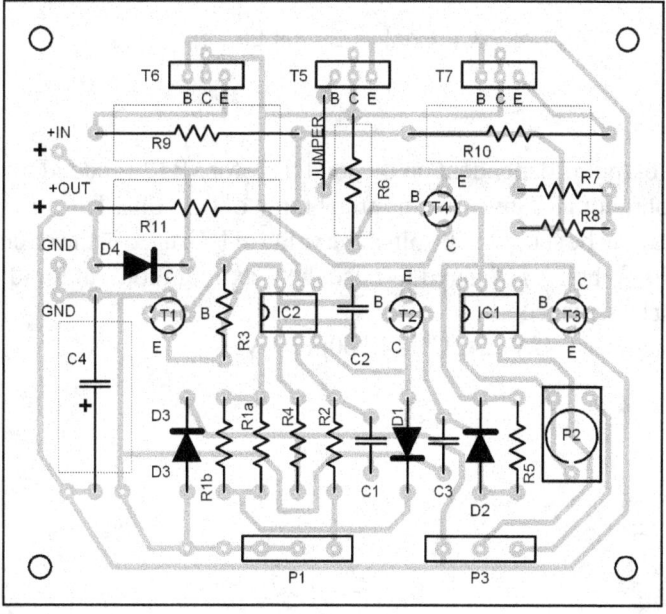

Figure 24.0 Parts Placement Layout

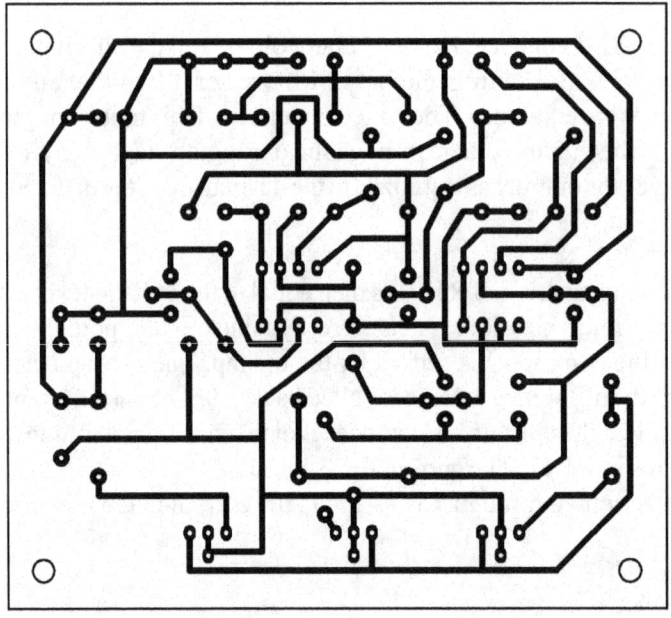

Figure 24.1 Printed Circuit Layout

The minimum output is dependent on the load therefore R12 is added to the circuit as a fixed load. Using 470 ohm for R12 gives a minimum output of 0.4 volts. The maximum output voltage is fixed by R1b and can be set up to 50 volts. The value of R1b used in the circuit sets the maximum output to 45 volts. When it is desired to raise the maximum output to 50 volts, the following components must be replaced:

Transformer secondary 42V/2A, C5= 4,700µF/80V.

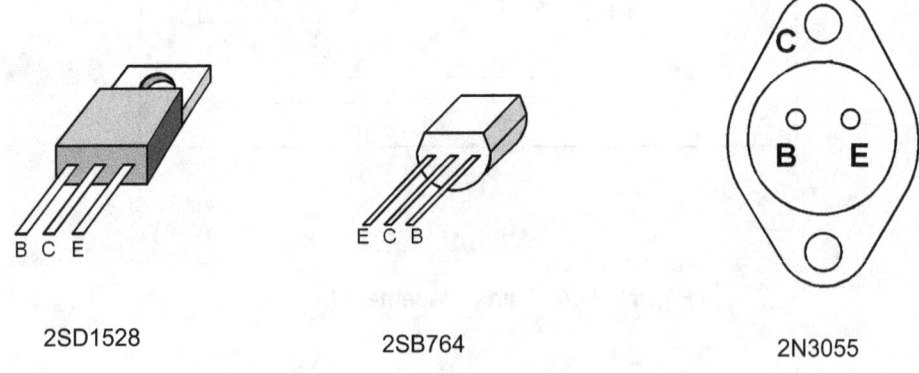

2SD1528   2SB764   2N3055

25 SUPPLY FOR OPAMPS

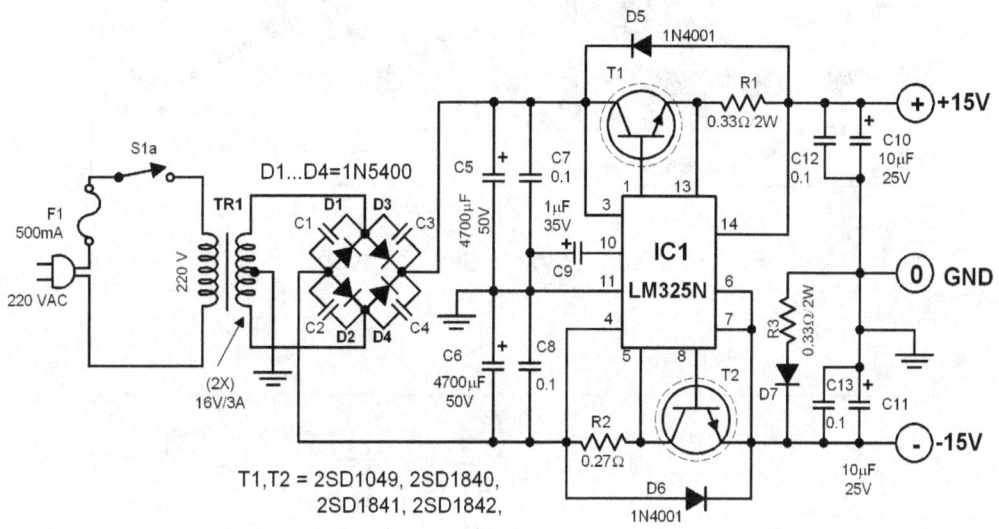

Diagram 25.0 Supply for OPAMPS

This power supply circuit is designed to support a large number of opamp ICs like in mixer circuits and sound synthesizers. The regulation is done by the IC LM325. The output voltage is +/- 15 volts symmetrical. The output current can be up to 2 amperes. Capacitors C9, C10, and C11 are tantalum types. The two power transistors T1 and T2 must be properly heatsinked.

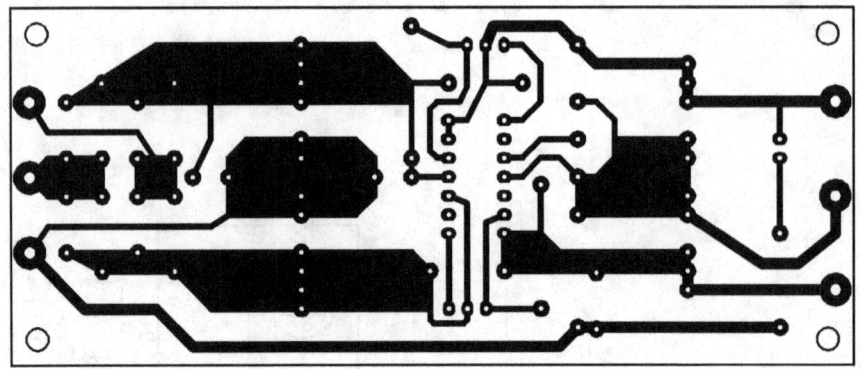

Figure 25.0 Printed Circuit Layout

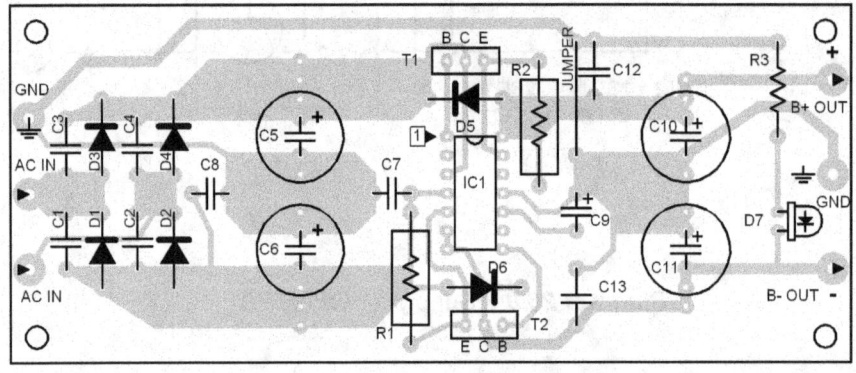

Figure 25.1 Parts Placement Layout

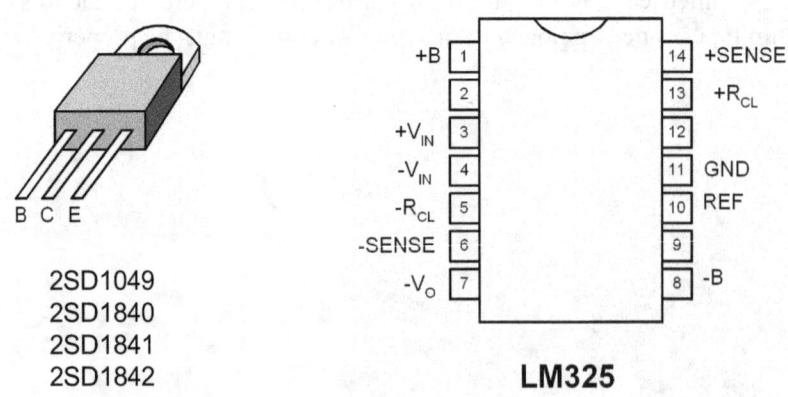

2SD1049
2SD1840
2SD1841
2SD1842

26 PS w/ DISSIPATION LIMITER

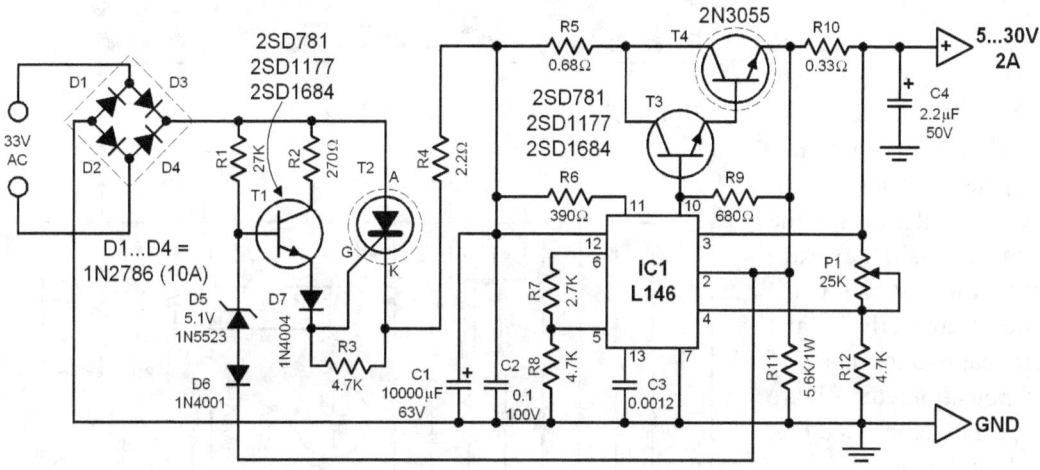

Diagram 26.0 PS with Dissipation Limiter

An ordinary power supply normally suffers from high dissipation levels. It happens when the output voltage is set at low level while the input level remains at maximum. The resulting difference voltage then "fries" the regulator circuit's power transistor continuously. This shortens the life expectancy of the transistor. Also, the heat build up in such regulators is very high that an internal air blower is almost always needed.

The power supply featured here does not know this kind of problem since it has an automatic dissipation limiter. The actual limiter circuit is composed of the components T1, T2, D5, D6, D7 between the transformer and the regulator stages. The triac T2 must be selected properly to handle the maximum current and dissipated power that will load on it. You can use the table at the end pages of this book to select the needed triac.

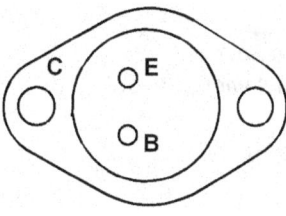

Bottom view
2N3055

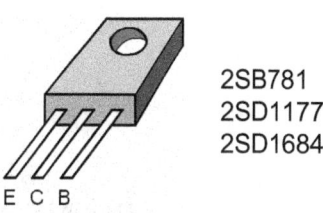

2SB781
2SD1177
2SD1684

E C B

The potentiometer P1 in the regulator stage is the normal output voltage adjuster. The output can be varied from 5 up to 50 volts.

The triac T2 and the power transistor T4 are to be installed on heatsinks outside the circuit board. The potentiometer P1 must also be installed outside the circuit board. Make sure to wire these external components correctly. Take note that two terminals of the potentiometer P1 are shorted together. See Fig. 26.2 for details.

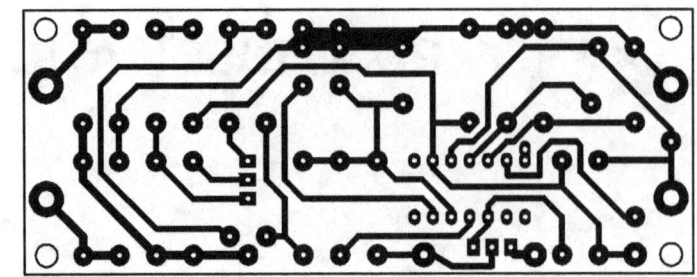

Figure 26.0   Printed Circuit Layout

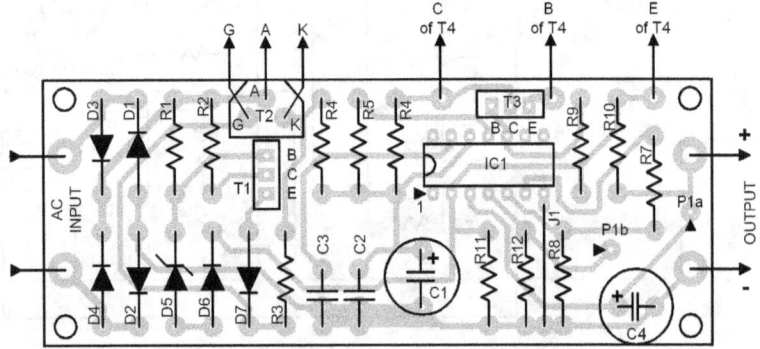

Figure 26.1   Parts Placement Layout

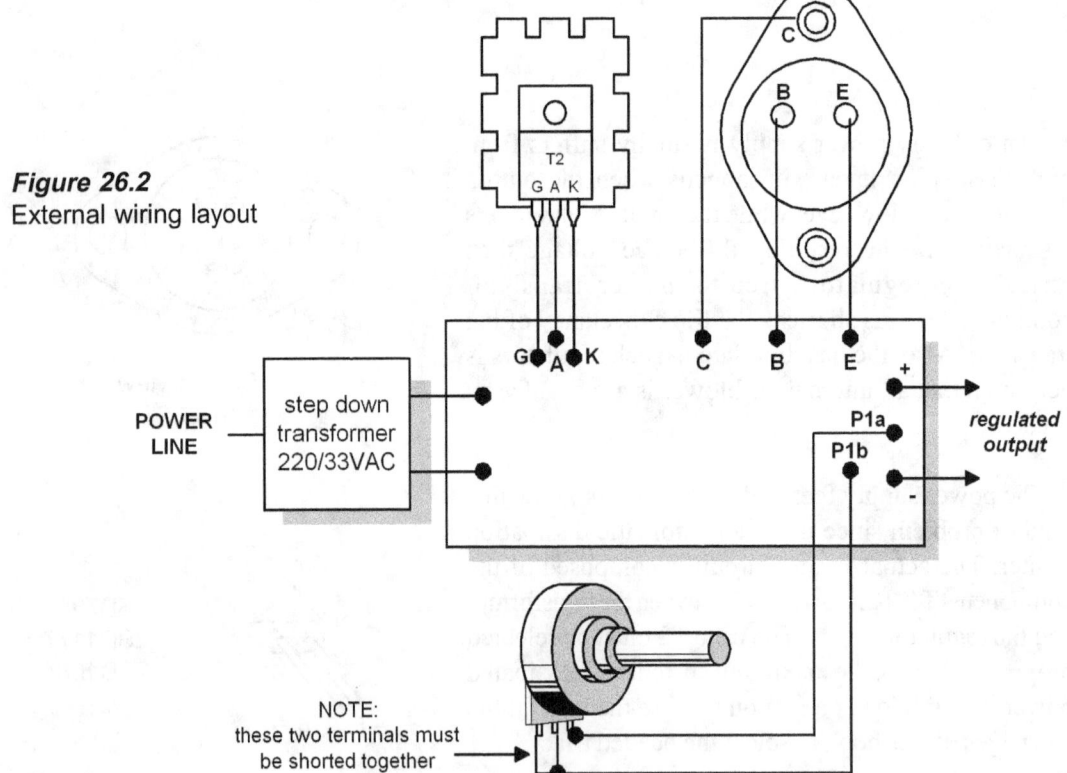

Figure 26.2
External wiring layout

27   5A DELAYED POWER-ON

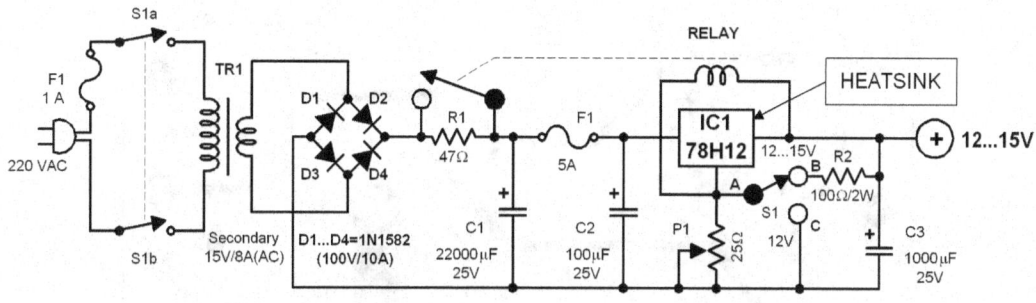

Diagram 27.0   5A Delayed Power-On

This power supply has a short power-on delay. This short delay has two advantages: First, the initial short high current present during switching a power supply with high value capacitors and high efficient transformers is avoided. Second, this short delay prevents the over-voltage fluctuations produced during switching from reaching the load circuits. This is very important for load circuits which are highly sensitive to voltage spikes.

The regulator can be operated in either fixed output mode or variable output mode. Switch S1 selects between the two modes. Position B is the fixed output mode and the power supply delivers a well regulated 12V. Position C is the variable mode and the output can be varied from 12 volts to 15 volts. The output voltage can be varied through P1. The maximum current output of 78H12 is 5A and the maximum short circuit current is around 7A. The IC must be heatsinked properly.

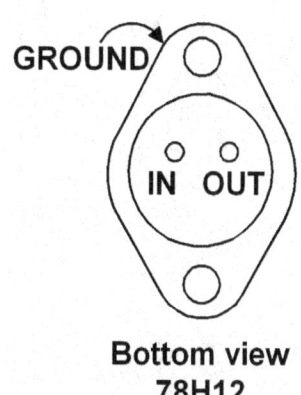

GROUND

IN   OUT

**Bottom view
78H12**

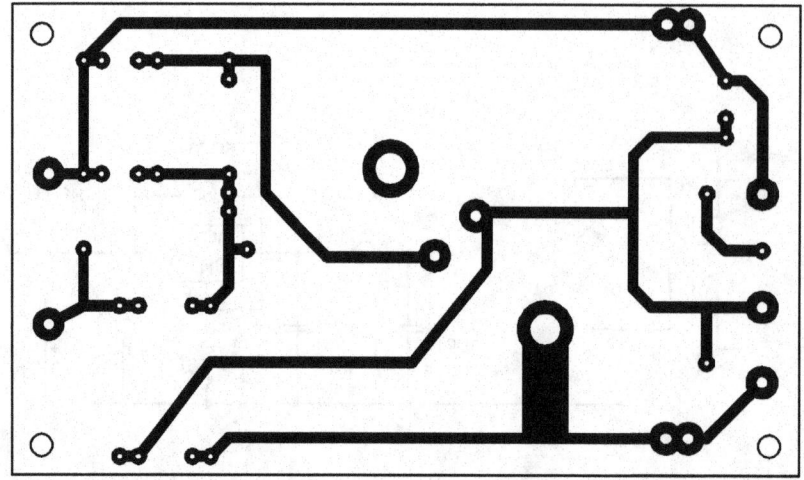

Figure 27.0   Printed Circuit Layout

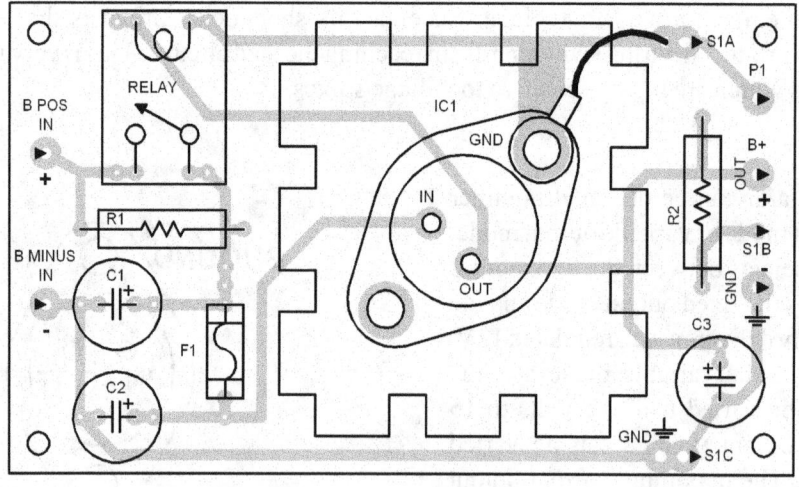

Figure 27.1   Parts Placement Layout

28 TTL POWER SUPPLY MONITOR

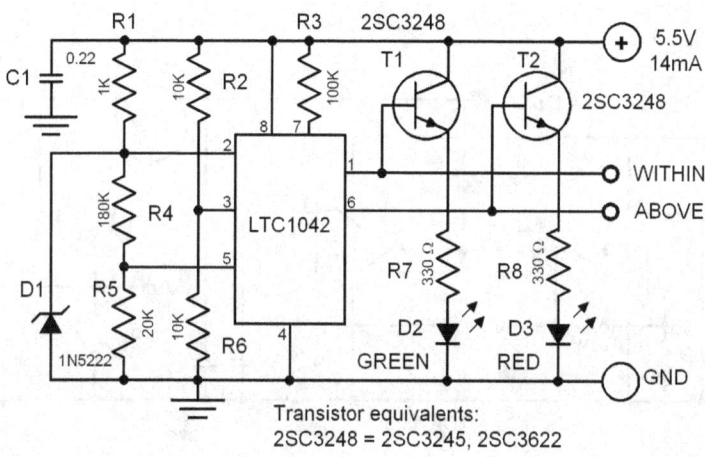

Diagram 28.0 TTL Power Supply Monitor

This simple to construct circuit monitors the 5-volt level TTL power line and gives a signal whether the supply voltage is outside or within the necessary range or "window". The heart of the circuit is a low-current integrated window comparator. The center of this window is set at 2.5 volt +/- 0.005 volt by the band-gap reference diode D1 which is connected to pin 2 of LTC1042. The width of this window must be 20% (+/- 10%) of the reference voltage.

The reference voltage is reduced by 25% through resistors R4 and R5 and fed to pin 5 (width/2 input) of the IC. The monitored voltage is then fed to pin 2 (window center input) so that the green LED (D2) will light up when the voltage is within the desired range. Otherwise, the red LED (D3) will light up signalling that the voltage is out of range.

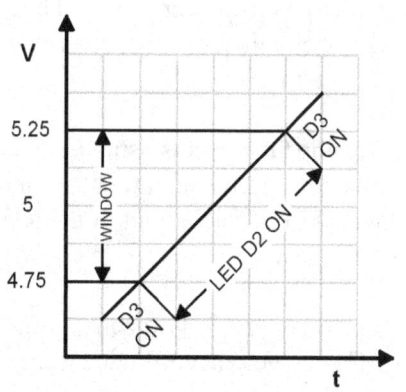

Table 28.0 Time vs Voltage

29   6V to 12V CONVERTER

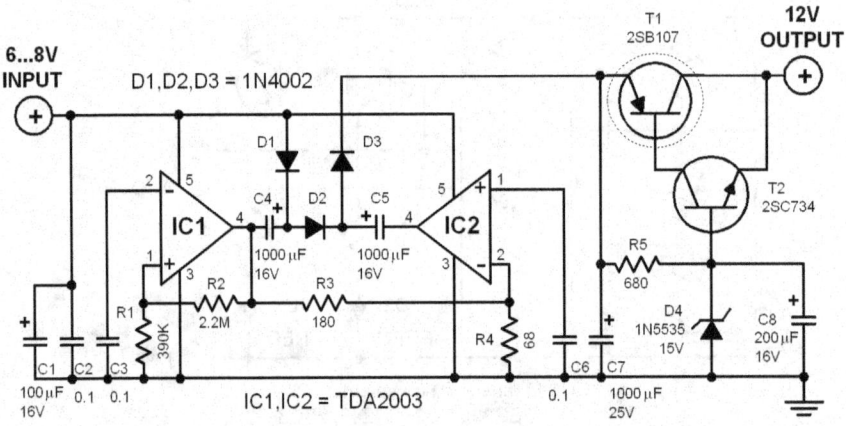

Diagram 29.0   6V to 12V Converter

In some situations, one needs a 12 volts power supply but has only a 6 volts battery available. In such a case, a voltage converter such as this featured circuit comes very handy. This circuit is a simple converter made with an IC from SGS with several additional external components. The IC is a TDA2003 but it can be replaced with a TDA2002. The cost of building the circuit should be low enough to justify constructing it instead of modifying the entire equipment setup to work directly with a 6 volts power supply. The two principles of simplicity and low cost are applied in this electronic circuit. It only uses two low cost AF amplifiers and functions properly without the need for a transformer.

The IC1 opamp functions as a stable power multivibrator. Its oscillation frequency is determined by the capacitor C3. Its oscillates at around 4 kHz at standby and increases in a loaded condition up to around 7 kHz. The output of the IC2 opamp is identical to the IC1 oscillator signal but in the opposite phase.

When the output of IC1 is at zero (saturation point of the output transistor to the ground), the capacitor C4 charges via the diode D1 up to the power supply level minus the voltage drop at D1. When the IC1 swings to the opposite direction, its output becomes positive. The output voltage from IC1 adds up to the voltage stored at the capacitor C4 forcing the diode D1 to stop conducting. Capacitor C5 then charges via the diode D2 to a voltage that is double than the power supply level. Due to the opposite phase of IC2, the negative pole of C5 is connected to the ground through the IC2 at this moment. By the next swing of IC1, its output is again at zero potential. At this moment the capacitor C4 is recharged while at the same time the voltage at C5 is „lifted" up to the voltage level present at IC2.

Capacitor C5 delivers its voltage to the output capacitor C7 via the diode D3. Theoretically, the output voltage has been multiplied by three. In practical application, however, the output voltage available from the capacitor C7 is lower than the theoretical value. This voltage level is also dependent on the load. Data gathered from testing the circuit showed that a 6 volts supply (from a battery) produces a voltage output of around 18 volts with no load to the circuit. This level sinks to about 12 volts when a load is applied that consumes a current of around 750 milliamperes. When the load current is about 400 mA, the voltage at the capacitor C7 is about 14 volts. These values are enough to power a standard car radio.

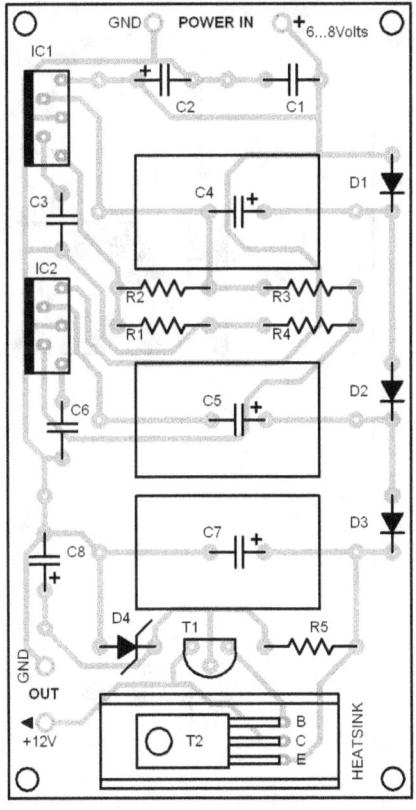

Figure 29.0 Parts Placement Layout

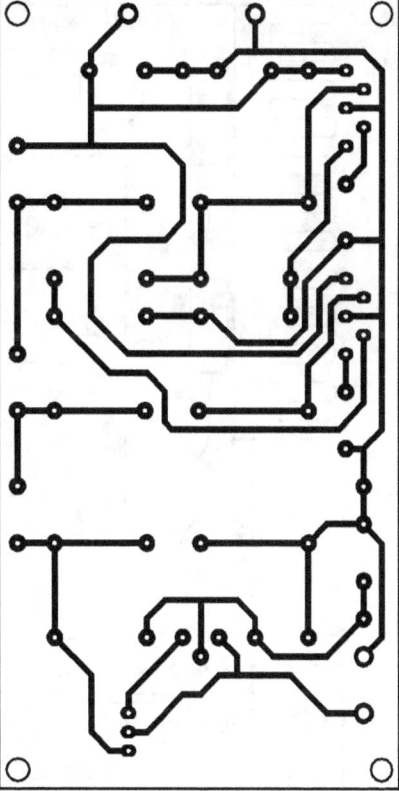

Figure 29.1 Printed Circuit Layout

As mentioned in the preceding page, the theoretical output could reach the triple of the supply voltage. To guard against unnecessary voltage increases at low current consumptions, a limiter stage was added to the circuit composed of a 15 volt zener diode and a darlington transistor T1/T2. This stage caps the output voltage to about 14.2 volts. To filter out ripple from the output, the capacitor C8 was also added. This helps prevent the hum signal from being noticed on radio or audio devices.

In constructing the circuit, attach the ICs to a common heatsink close to the pcb. The transistor must be attached to a separate heatsink. The two ICs have built-in protection against short circuits and thermal overload. The TDA2002 can be used in place of TDA2003. The TDA2003 however, has better characteristics. To get a much higher current output from the circuit, the capacitors C4, C5 and C6 must be increased to 2200 μF.

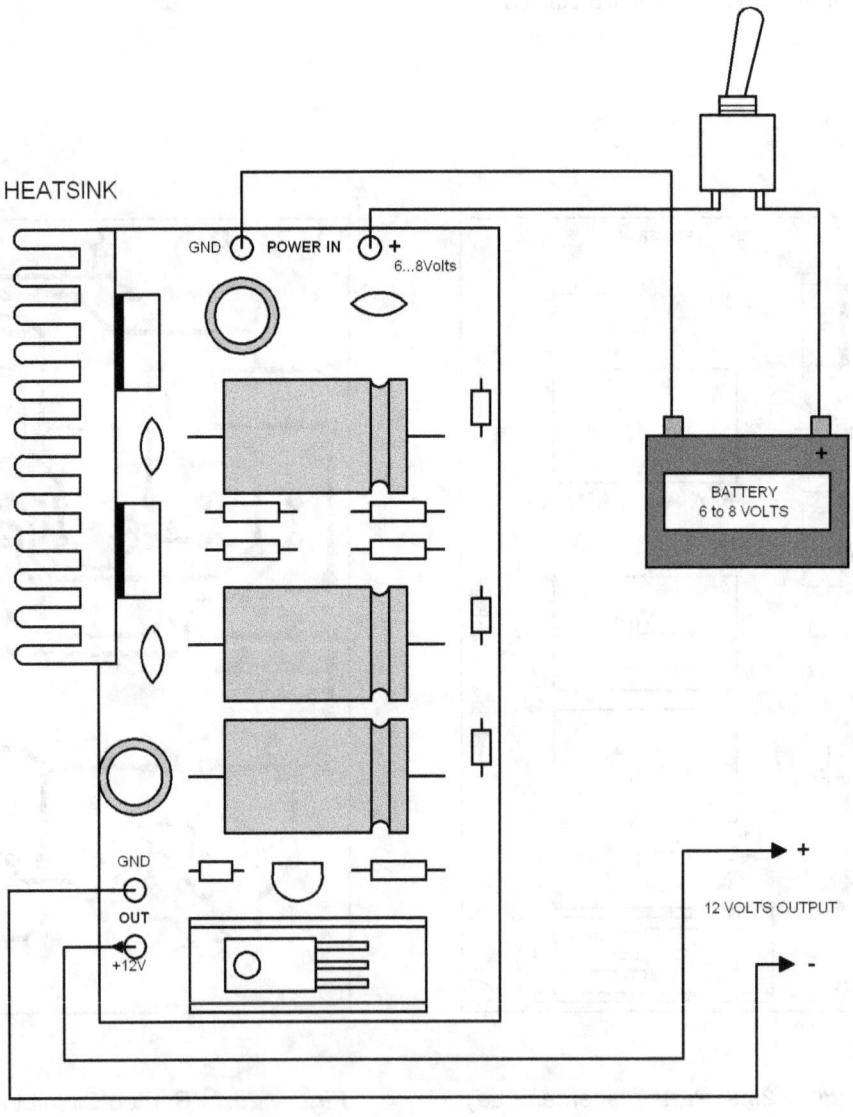

Figure 29.2 External Wiring Layout

30 VERSATILE POWER SUPPLY

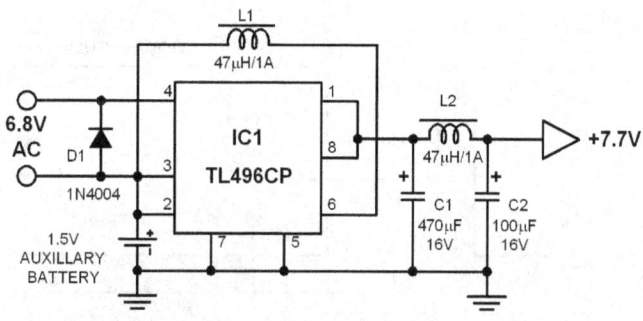

Diagram 30.0 Versatile Power Supply

This circuit is called versatile power supply. It delivers 7.7 volts output converted from either a single 1.5 volt battery or from a 6.8V AC transformer. The output is higher than the input voltage. The voltage increase is done by the integrated chip TL496CP. Not only that, both an AC input and a battery can be connected simultaneously as the power source. The AC input is controlled by a series regulator while the battery is controlled by a switching regulator.

As long as the series regulator delivers the necessary output, the switching regulator remains disabled. When a NiCad cell is used as a battery, it will be charged simultaneously by the AC input. Once the AC input is disconnected or when the AC line goes off, the battery automatically takes over the function of supplying the needed 7.7 volts. A really versatile power supply.

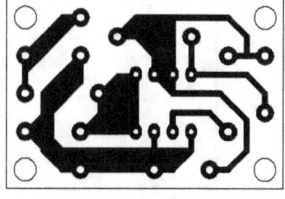

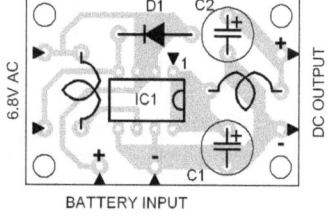

Figure 30.0
Printed Circuit Layout

Figure 30.1
Parts Placement Layout

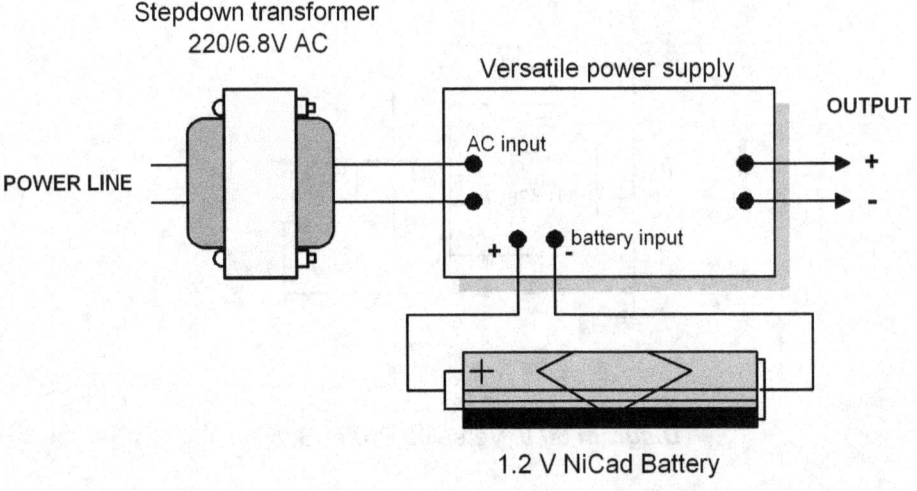

Figure 30.2 External wiring layout for the versatile power supply

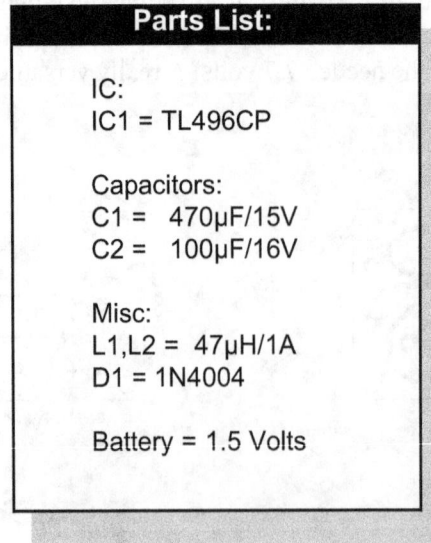

Parts List:

IC:
IC1 = TL496CP

Capacitors:
C1 = 470µF/15V
C2 = 100µF/16V

Misc:
L1,L2 = 47µH/1A
D1 = 1N4004

Battery = 1.5 Volts

31 ROBUST 5V SUPPLY

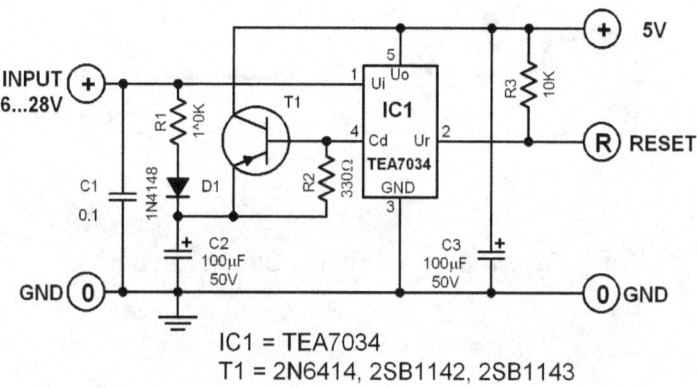

IC1 = TEA7034
T1 = 2N6414, 2SB1142, 2SB1143

Diagram 31.0 Robust 5V Supply

This 5V supply circuit is very durable and insensitive to voltage fluctuations. It can block voltage spikes up to 80V. Very short current interference does not affect the function of the IC. This circuit is specially designed to support microprocessor circuits.

The maximum current output is 500 mA. The voltage drop between the input and the output voltages is around 0.6V. The output voltage is 5V+/-2.5%. The IC is also protected from thermal overloads.

Parts List:

Resistors:
R1,R3 = 10K
R2 = 330Ω
Capacitors:
C1 = 0.1/50V
C2,C3 = 100µF/50V

Diode:
D1 = 1N4148
IC:
IC1= TEA7034

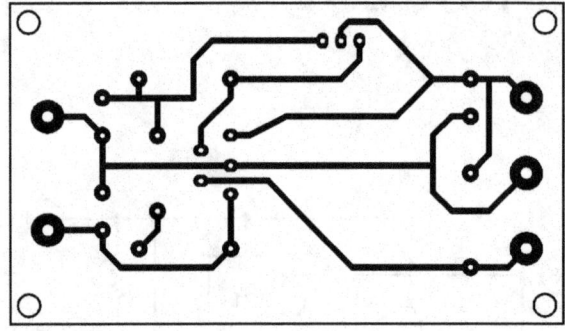

Figure 31.0   Printed Circuit Layout

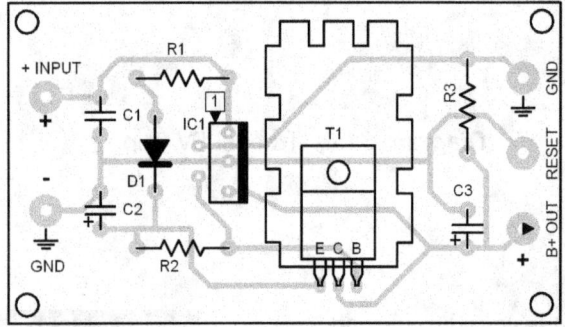

Figure 31.1   Parts Placement Layout

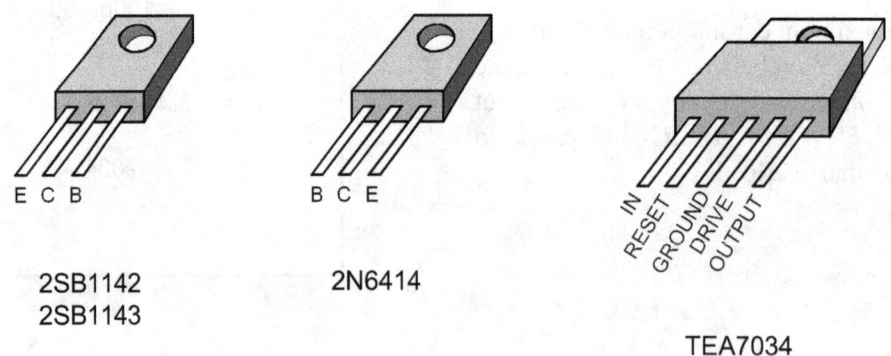

2SB1142
2SB1143

2N6414

TEA7034

32   VOLTAGE CONVERTER

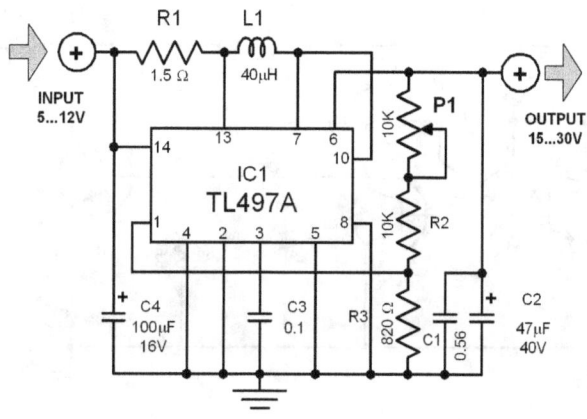

Diagram 32.0   Voltage Converter

This circuit converts an input voltage of 5...12 volts to a higher level of 15...30 volts. This is specially helpful in mobile applications where power supply levels are commonly limited to 12 volts supplied by batteries.

The circuit uses the IC L4973 as a flyback-voltage converter. The choke coil is 40µH/2A. Capacitors C2 and C1 suppress voltage spikes. The maximum output current depends on the difference between the input and output voltages. This maximum is around 100 mA. The ripple voltage is relatively low. The standby current is around 8 mA and the efficiency is about 70%.

The maximum output current is 100mA!

33 FUSE MONITOR

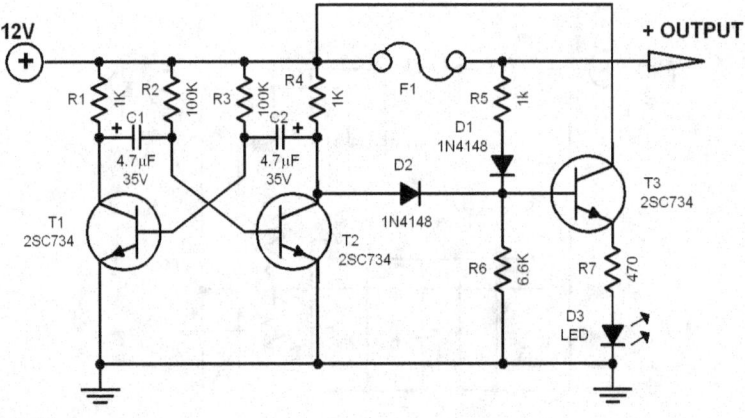

Diagram 33.0 Fuse Monitor

This circuit monitors a DC fuse. Its LED lights continuously when the fuse is intact but blinks if the fuse is busted. The featured circuit is designed for 12 volts but can be modified for other voltages. To use the circuit for 6 volts, divide all resistance values by two. For 24 volts, double all resistance values.

The circuit is made of an astable multivibrator (AMV) and a LED driver. The biggest part of the circuit is before the fuse with the exception of R5. The AMV functions all the time as long as the power is on. The output of AMV is connected to the driver circuit via D2. As long as the fuse is intact, current flows to the base of T3 via R3 and D1 and the LED lights. Once the fuse gets blown, no more current can flow to the base of T3. In this case, the AMV takes over the control of the LED driver and the LED blinks.

The circuit consumes around 25 mA. Most of the current is consumed by the LED. If one decides to use the circuit in battery operated modules, it is highly recommended to use a high efficiency LED and increase the value of R7 accordingly.

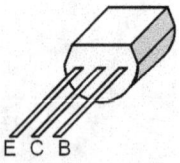

E C B

2SC734

34 LOW DROP REGULATOR

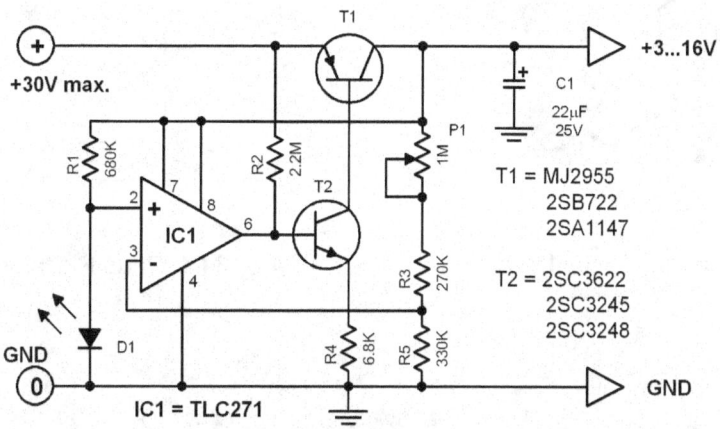

Diagram 34.0 Low Drop Regulator

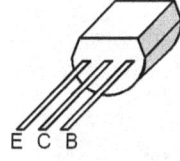

2SC3622
2SC3245
2SC3248

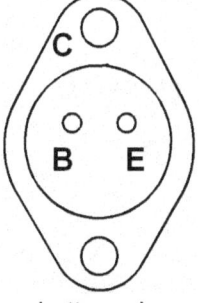

bottom view

MJ2955
2SB722
2SA1147

This power supply regulator has the following characteristics:

a) very low voltage drop of less than 1V.
b) negligible standby current from 20 to 30 µA.
c) variable but very stable output.

These characteristics make the circuit practical in some applications compared with the 3-terminal IC regulator which has high voltage drop and standby current consumption. Specially in battery operated applications, the discrete regulator featured here is highly recommended.

The circuit is basically a series regulator that uses a normal LED to produce the reference voltage. The given component values deliver a variable output from 2V up to 8V. If you intend to increase it up to 16V, change R4 with 220 kiloohms. Eventually, you need to increase the value of P1. The maximum current that can be delivered by the regulator circuit depends on the type of the series transistor used and the difference between the input and output voltages.

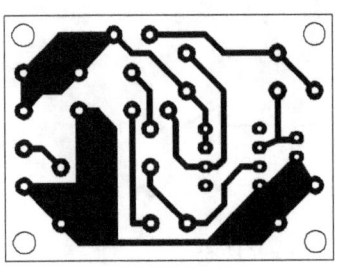

Figure 34.0 Printed Circuit Layout

Figure 34.1 Parts Placement Layout

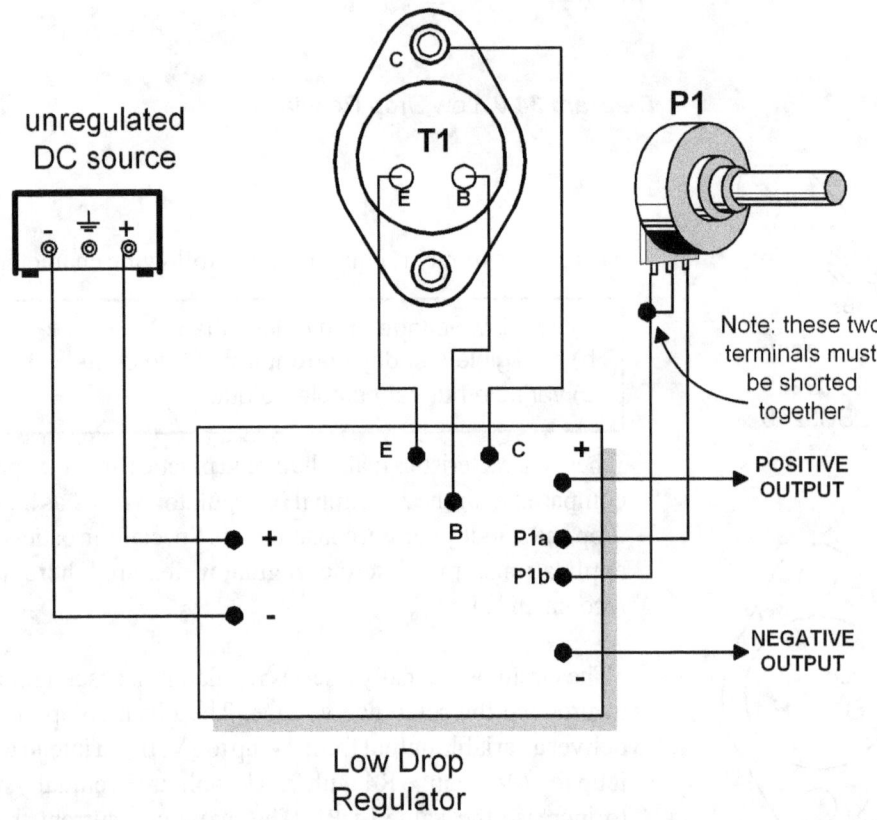

Figure 34.2 External wiring layout of the low drop regulator. Take note that two terminals of the potentiometer P1 must be shorted together. Transistor T1 must be properly heatsinked.

35 AMPLIFIED REGULATOR

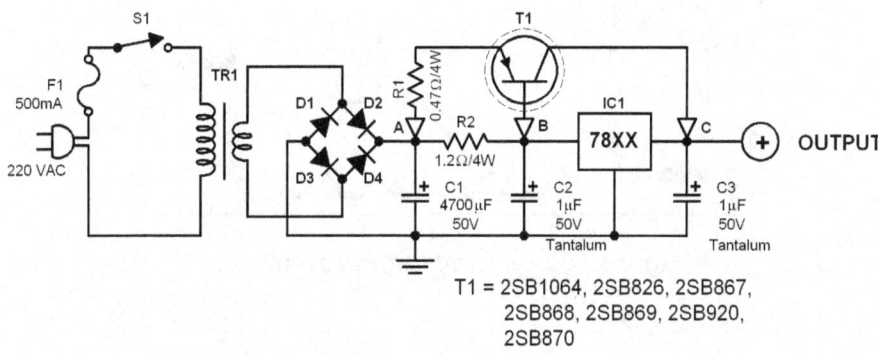

Diagram 35.0 Amplified Regulator

Normally, 78xx voltage regulators can only deliver up to 1 ampere. This circuit can deliver currents up to 10 amperes by adding a current amplifier to the common 78xx series voltage regulators. The current amplifier can be a single or two power transistors in parallel depending on the needed maximum current output.

Table 35.0 shows the necessary component values for every maximum current output. The transistors must be heatsinked to minimize voltage drop caused by increasing temperature. When two transistors are used in parallel (additional $T1°$ transistor as shown in Diagram 64.1), a resistor $R1°$ must be added to each emitter terminal (see Diagram 35.1). The PCB layout for the higher current version is shown in figures 35.2 and 35.3. The transistor must be installed in a separate heatsink.

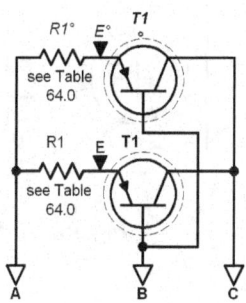

Diagram 35.1 Additional T1($T1°$)

78xx
3-Terminal
voltage regulator

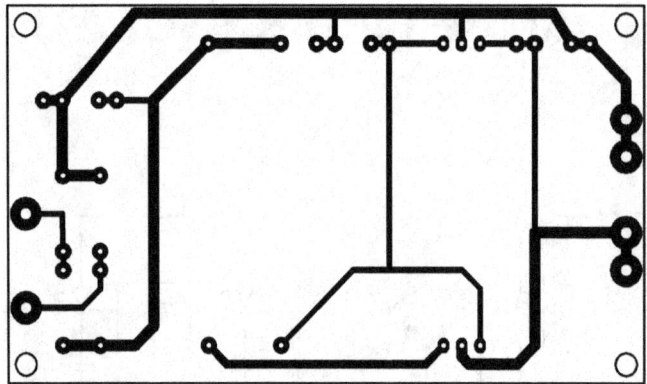

Figure 35.0 Printed Circuit Layout

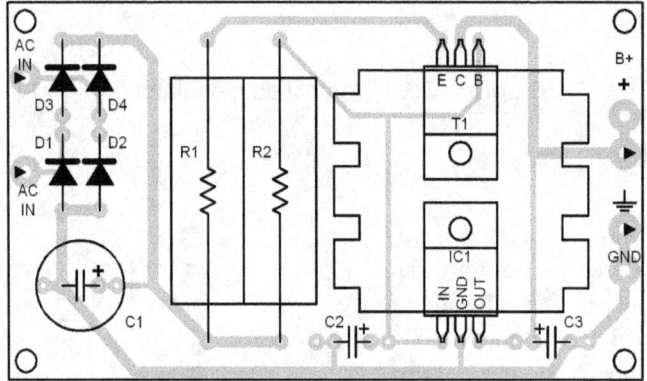

Figure 35.1 Parts Placement Layout

Maximum Current	D1...D4	C1 in μF	R1(*R1°*) in Ω/4W	R2 in Ω/4W	T1	*T1°*
2A	1N5401	4700 or 2 x 2200	0.47	1.2	2SB1064	
3 A	ECG5863	6800 or 3 x 2200	0.39	2.2	2SB638	
4 A	6...8A	10000 or 2 x 4700 or 4 x 2200	0.27	2.2	2SB638	
5 A	1N1582 8...10A	10000 or 2 x 4700 or 4 x 2200	0.22	2.2	2SB638	
7 A	10...14A	15000 or 3 x 4700	0.27	2.2	2SB638	2SB638
10 A	15...20A	22000 or 2 x 10000 or 4 x 4700	0.18	2.2	2SB638	2SB638
					See Diagram 25.0 for transistor equivalents.	

Table 35.0 Component values

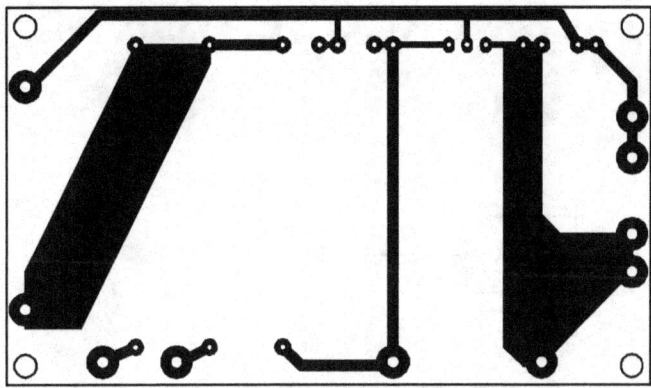

Figure 35.2 Printed Circuit Layout for
the higher current version

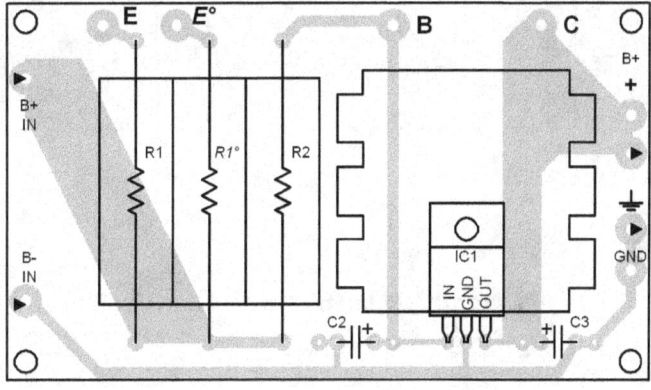

Figure 35.3 Parts Placement Layout for
the higher current version

B C E

2SB1064, 2SB826, 2SB867,
2SB868, 2SB869, 2SB920,
2SB870

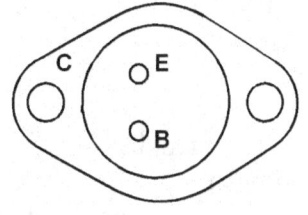

2SB638 2N6285
2SB639 2N6286

36 78XX REGULATOR MONITOR

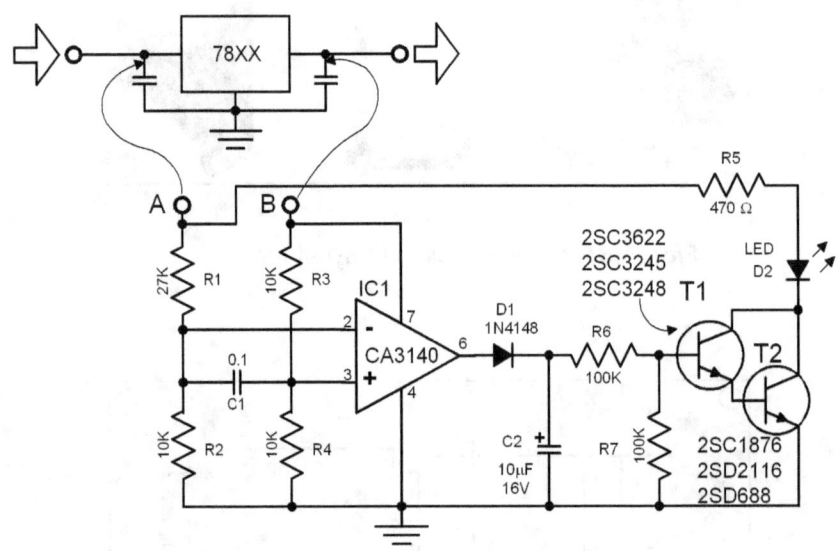

Diagram 36.0 78XX Regulator Monitor

Although 78XX 3-terminal regulators are known to be stable, problems sometimes arise in their applications. The input voltage of these ICs must be at least 3 volts higher than the output voltage. But sometimes these ICs breakdown due to overloads or internal defects. The monitor circuit featured here continuously monitors a 78XX IC regulator whether if functions according to its specifications or not. The value of R1 in the circuit is computed for a 7805 regulator. If you use the monitor for other regulator IC types, change the R1 value by using the formula shown on the following page:

The input and output terminals of the regulator are connected to the inputs of IC1 which is configured as a difference amplifier. When the input voltage of the 78XX IC is low, the output of IC1 increases enough to charge C2 and turn on T1. In this case, the LED lights up. C2 holds the LED for about 10 mS longer before it finally turns off so that even short voltage collapses can be registered.

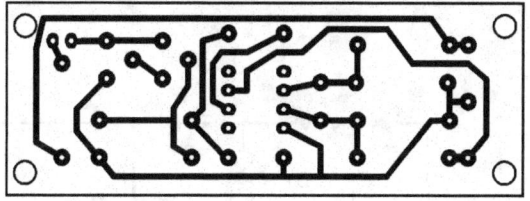

Figure 36.0 Printed Circuit Layout for
the 78XX Regulator Monitor

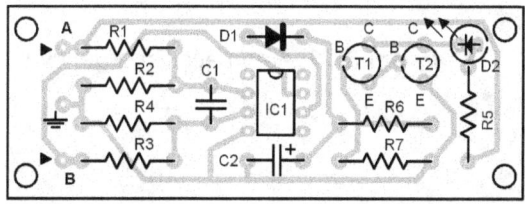

Figure 36.1 Parts Placement Layout

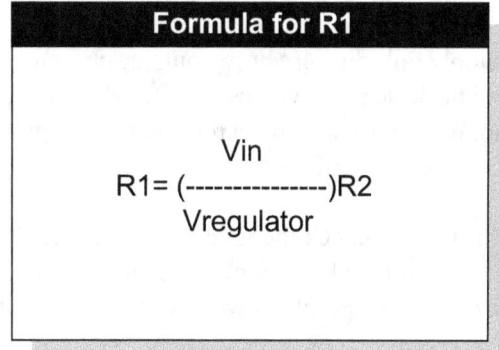

Formula for R1

$$R1 = \left(\frac{Vin}{Vregulator}\right) R2$$

37 EMERGENCY LAMP

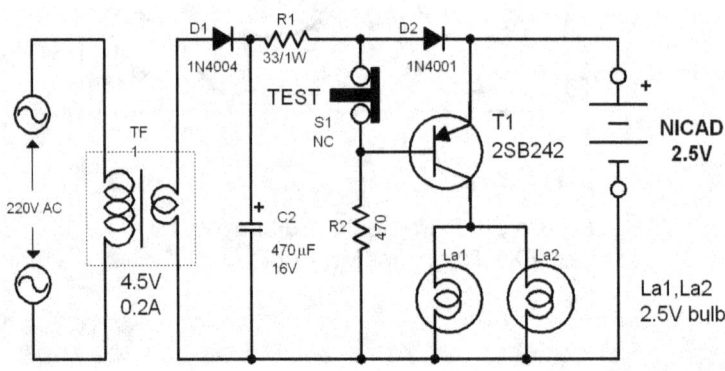

Diagram 37.0 Automatic Emergency Lamp

This electronic circuit automatically turns on a lamp during brownouts. Its power source is a NiCad battery that is being charged continuously by the main power line when there is no brownout. The diagram shows the simplicity of the circuit.

The electronic circuit works this way: the voltage from the step down transformer is rectified by diode D1 and filtered by C2. The NiCad battery gets charged with about 100 mA via the diode D2 and resistor R1. The NiCad battery must have a capacity of at least 2 Ah to tolerate the charging rate. In a normal situation, the base voltage of T1 is positive in relation to its emitter due to the voltage drop at diode D2. Transistor T1 in this case does not conduct and the lamps stay unlit.

When the main power supply fails during a brownout, the charging current is interrupted. In this case, a current flows from the base of T1 via resistor R2 which triggers the transistor to conduct and the two lamps light up. When the main power returns, the charging current flows again through D2 and the transistor turns the lamps off.

The „TEST" button S1 is used to check the function of the circuit. If the secondary coil of the transformer delivers a higher voltage level, replace the R1 with a higher value resistor to avoid exceeding the maximum charging current allowed for the NiCad battery being used.

38   OVERVOLTAGE CROWBAR

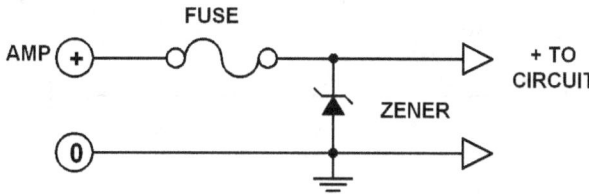

Diagram 38.0   Overvoltage Crowbar

Simple but very effective electronic circuitry. This very simple circuit can protect a main circuit which is sensitive to overvoltages. The first circuit uses a zener diode in parallel to the power line of the protected circuit or device. Once the output of the power supply increases due to a malfunction, the excess voltage will be routed to ground by the zener diode. If the voltage will increase further, the total current consumption will exceed the capacity of the fuse. The fuse will blow at this moment and the current supply is broken.

The zener voltage of the zener diode must be 1 or 2 volts higher than the correct supply voltage and must have a higher current rating than the fuse. Of course the maximum allowable supply voltage of the protected circuit must be taken into account. For example if the supply voltage must be 15 volts and the maximum allowable voltage of the protected circuit is 18 volts, then a zener diode of 18 volts must be used.

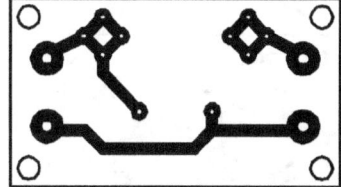

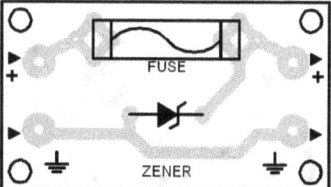

Figure 38.0.0   Printed Circuit Layout   ***Figure 38.0.1***   Parts Placement

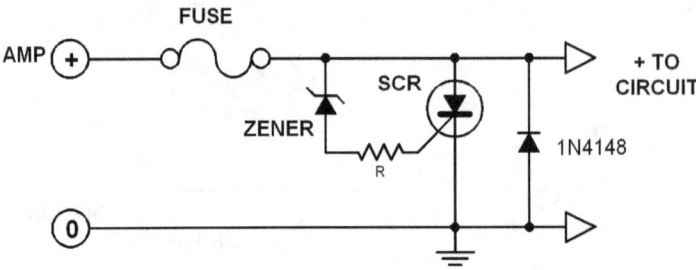

Diagram 38.1   Overvoltage Crowbar Nr. 2

The second circuit can be used for higher current ratings. The low current zener diode triggers the SCR when the supply voltage exceeds the safe level. AS a result, the SCR shorts the fuse to ground thereby blowing it. The resistor R limits the trigger current and the zener current flowing through the diode. The approximate value of R can be found by using this simple formula:

$$R = \frac{\text{Zener diode's voltage - SCR's trigger voltage}}{\text{SCR's trigger current}}$$

Of course, you have to select the right SCR type for your particular application. Use the tables on the end pages of this book.

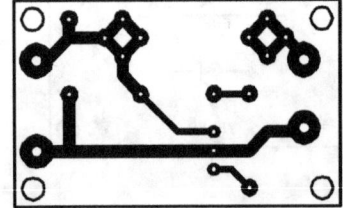

Figure 38.1.0   Printed Circuit Layout

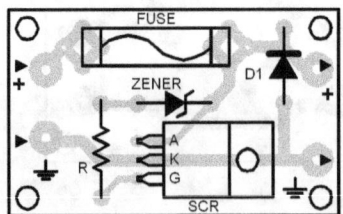

Figure 38.1.1   Parts Placement Layout

39 CONSTANT CURRENT

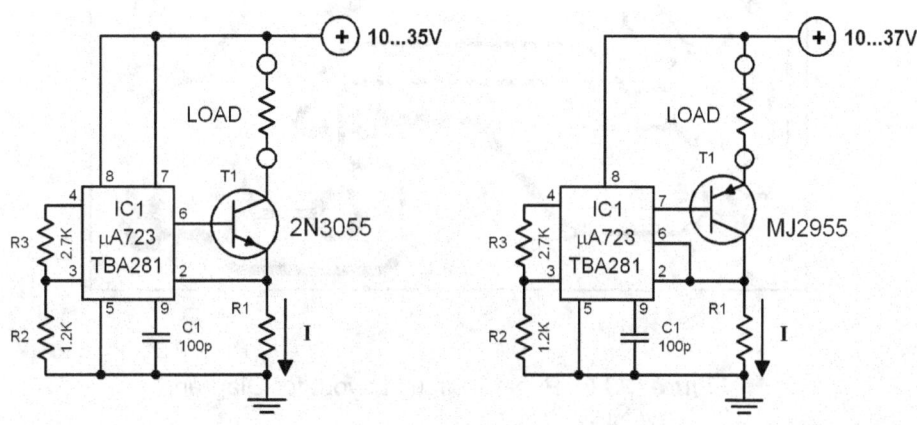

Diagram 39.0   Constant Current

There are different ways to construct a constant current source using a voltage regulator. This time we use the µA723. This IC is sometimes labeled as TBA281. The excellent stability and temperature coefficient of this IC is well known. It is actually designed as a voltage regulator but can be used to regulate current as well. Two circuit designs are featured here - one which uses a 2N3055 power transistor and another which uses a MJ2955 transistor. The constant current I can be found with this formula:

$$I = \frac{2.2V}{R1}$$

The load is connected to pin 7 of the IC. The maximum current load that can be handled safely by the IC is around 150 mA. Add a high power transistor to the circuit to increase it current handling capacity.

Circuit 1 of Diagram 39.0 uses an NPN transistor while circuit 2 uses a PNP transistor.

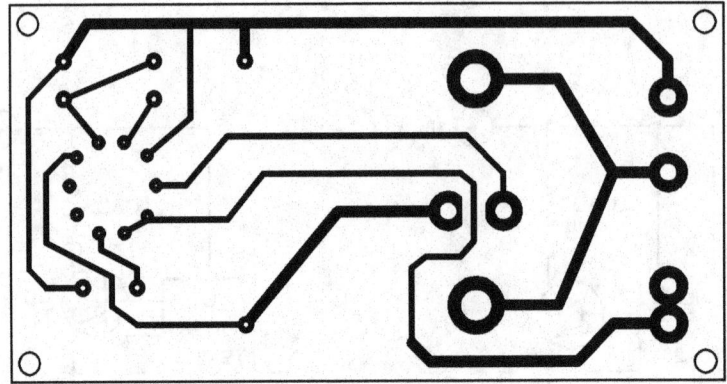

Figure 39.0 Printed Circuit Layout for Diagram
39.0 (w/ NPN transistor)

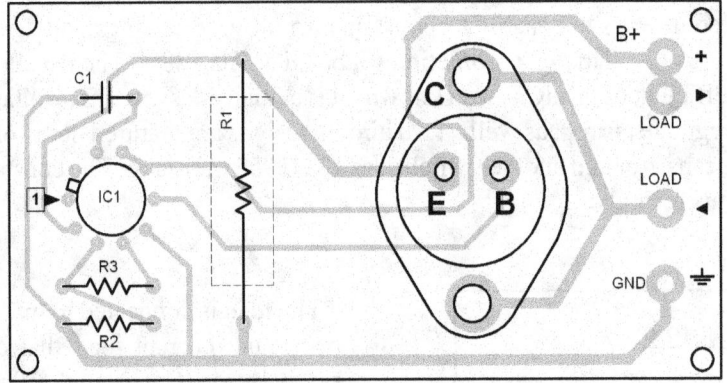

Figure 39.1 Parts Placement Layout for
Diagram 39.0 (w/ NPN transistor)

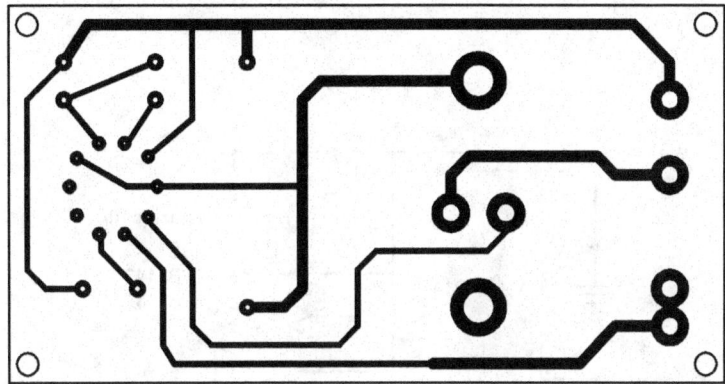

Figure 39.2 Printed Circuit Layout for Diagram 39.1 (w/ PNP transistor)

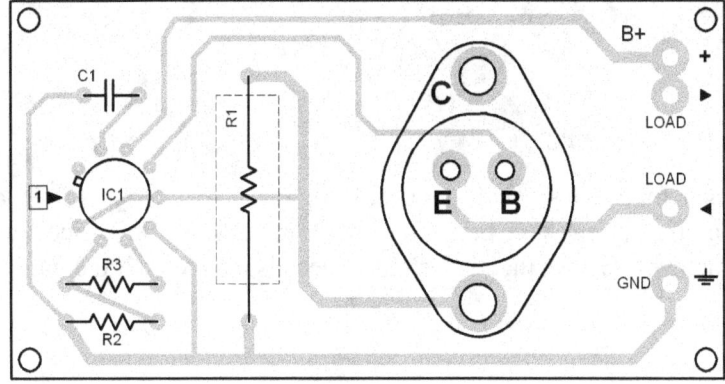

Figure 39.3 Parts Placement Layout for Diagram 39.1 (w/ PNP transistor)

40 DC MOTOR SPEED REGULATOR

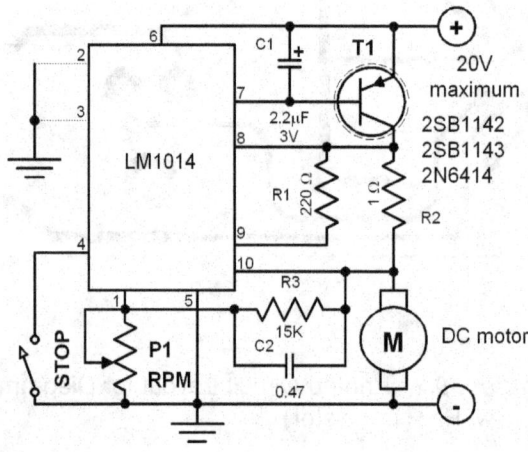

Diagram 40.0 DC Motor Speed Regulator

This speed regulator uses a single IC LM1014 to control the speed of a DC motor. It senses the increase in the motor-current when the rotation of the motor slows down due to a load. The IC then increases the motor voltage so that the original speed is recovered. Potentiometer P1 varies the speed of the motor.

41 SYMMETRICAL POWER SUPPLY

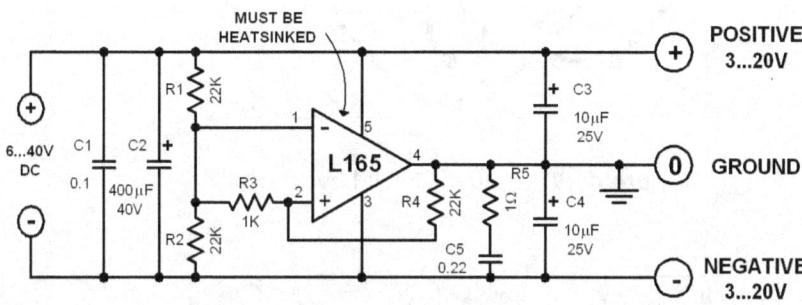

Diagram 41.0 Symmetrical Power Supply

The compact 5 pin L165 IC generates a stabilized symmetrical power supply from a single asymmetrical power supply. The output voltage is however, half of the input voltage. One only needs to add the ripple filter capacitors C1,C2, C3 and C4 and some resistors for setting the symmetry. In constructing the circuit, place the capacitors C1 and C2 as close to the IC as possible. On the other hand, place the capacitors C3 and C4 close to the output jacks. Make sure that the circuit lines on the pcb are properly dimensioned to handle high current levels. Current levels up to 3 amperes can flow through the circuit. Additionally, provide a proper heatsink for the L165 IC.

The IC can also be viewed as a voltage amplifier. It amplifies the voltage appearing at the junction between R1 and R2.

One can also replace the IC with TCA1365. However, when using the TCA IC, pins 3 and 4 have to connected together. Also, connect a 220 pF capacitor between pins 5 and 6.

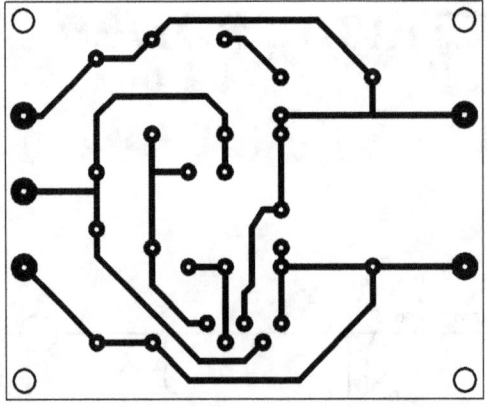

Figure 41.0 Printed Circuit Layout

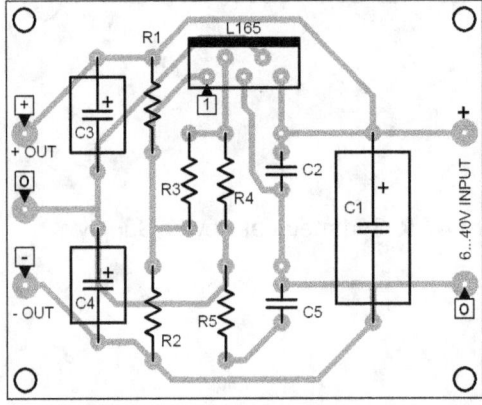

Figure 41.1 Parts Placement Layout

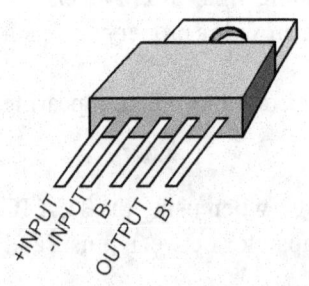

L165

Pin configuration

1 = + IN
2 = - IN
3 = B- (- VS case)
4 = OUT
5 = B+ (+ VS)

42   2N3055 DARLINGTONS

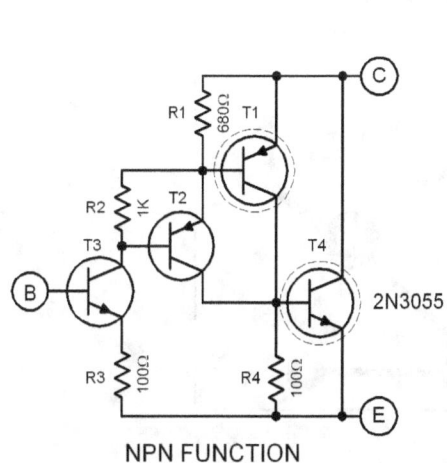

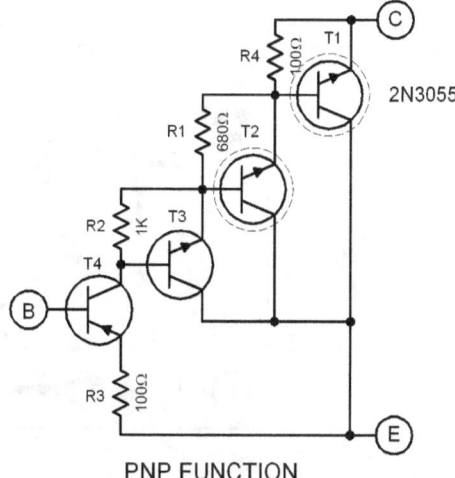

NPN FUNCTION   PNP FUNCTION

Transistors:
T1 = 2SB874, 2SB1144
T2 = 2SA1285, 2SA1285A
T3 = 2SC3245, 2SC3248

Transistors:
T2 = 2SD781, 2SD1177, 2SD1684
T3 = 2SC3245, 2SC3248
T4 = 2SA1285, 2SA1285A

Diagram 42.0   2N3055 Darlingtons

Very high capacity is the business of these circuits. The featured transistor combinations can be used for power supply regulators or final amplifiers that require high collector voltage, high current, high power dissipation capacity, and high current amplification characteristics. Both circuits use a 2N3055 transistor for the NPN and the PNP functions respectively. Using an NPN transistor for the PNP function may seem unusual if not incorrect to you at first glance. Through the special configuration, however, the second circuit functions as a PNP transistor. Just consider the whole circuit as a single high capacity transistor. In fact, the three solder points are labeled E,B,C. They correspond to the normal terminals of a single transistor.

The gain of the circuits (both AC and DC) is around 1.5 million. The maximum power dissipation at 25°C is 115 watts. The maximum collector voltage is 60 volts and the maximum collector current is 15 amperes. The voltage drop at the NPN combination is around 2 volts. The voltage drop at the the PNP combination is around 3 volts.

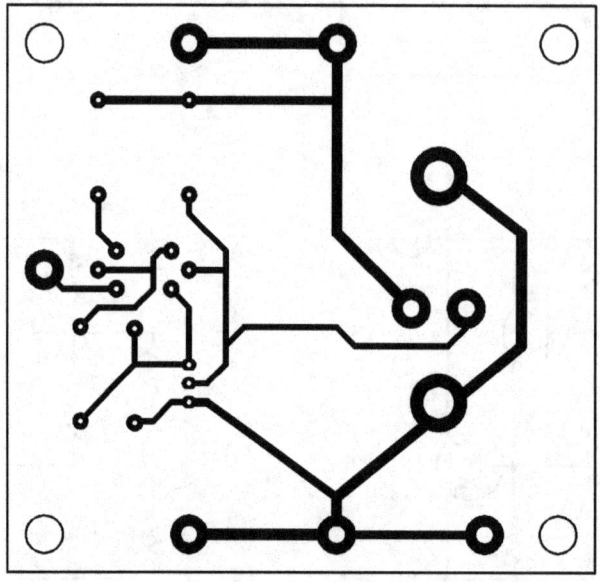

Figure 42.0 Printed Circuit Layout for the NPN function

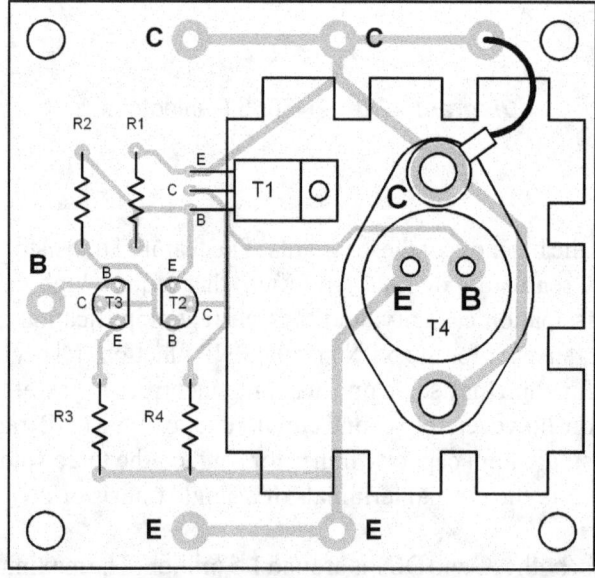

Figure 42.1 Parts Placement Layout for the NPN function

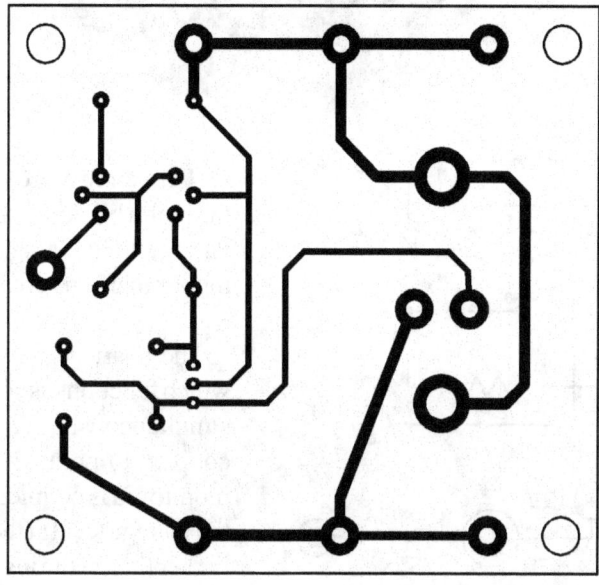

Figure 42.2 Printed Circuit Layout for the PNP function

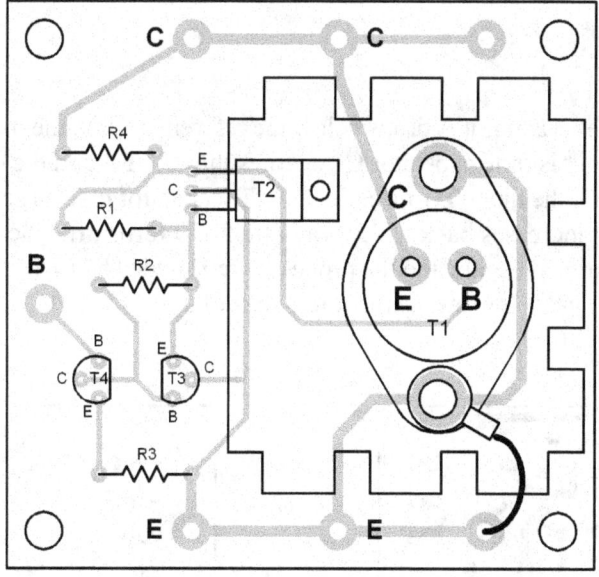

Figure 42.3 Parts Placement Layout for the PNP function

43 VOLTAGE MONITOR

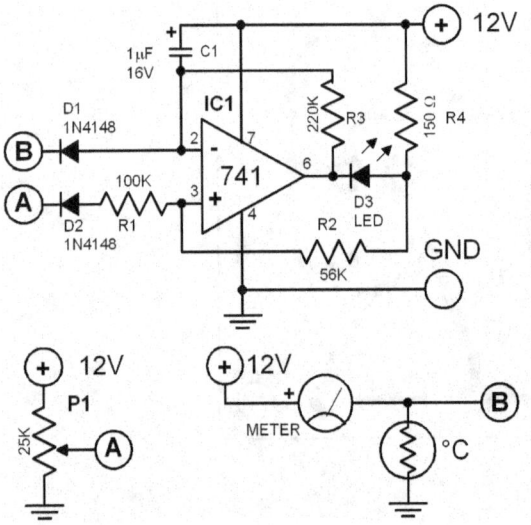

Diagram 43.0 Voltage Monitor

This circuit provides an optical signal through a blinking LED to show whether the voltage level being monitored is lower or higher than a reference level.

The heart of the circuit is a 741 opamp which functions as comparator and oscillator simultaneously. The reference voltage is coupled to point B and the voltage being monitored is coupled to point A. As long as the voltage level at the non-inverting input is higher than at the inverting input, the output of 741 is 12V and the LED is off.

Once the voltage level at point A drops below the reference level, the output of 741 becomes 0 and the LED lights. This time capacitor C1 charges through R3 and the output of the IC1. At a certain charging level, the diode D1 turns off and the capacitor discharges to the minus input of IC1. The IC1 output increases back to 12 V and the LED turns off. The capacitor continues to discharge until D1 turns on again and the process is repeated. The LED therefore blinks as long as the monitored voltage is lower than the reference voltage.

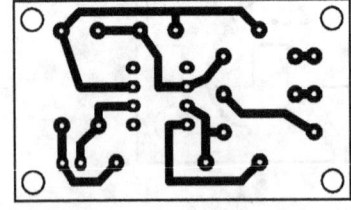

Figure 43.0 Printed Circuit Layout

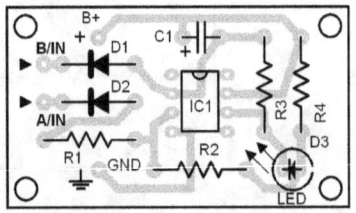

Figure 43.1 Parts Placement Layout

44 TEMPERATURE MONITOR

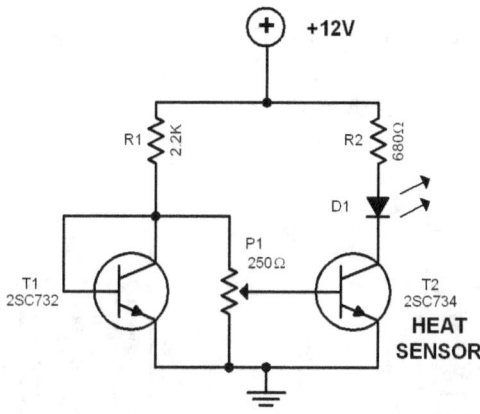

Diagram 44.0 Temperature Monitor

High power amplifiers usually dissipate large amount of heat. It is therefore practical to monitor the level of heat from its power transistors and heatsinks and if needed, to automatically turn off the amplifier to avoid damage to the vital components. This can be done by using simple temperature monitors such as the circuit featured here. Since it is not required to monitor the temperature by strict degree resolution, this simple monitor will work very well.

The electronic circuit works by comparing the voltage drop of a „cold" diode (T1) with the base-emitter voltage of a „warm" transistor (T2). The transistor must be attached physically closest to the heat source. Ideally it should be attached to the heatsink of the power transistors. The diode (T1) must be positioned away from the heat source to ensure that it is always at room temperature. The circuit „measures" the heat difference between the transistor and the diode. The diode (T2) is connected to the power supply via R1. The base of T2 is connected to the anode of the diode via the potentiometer P1. The LED D1 and resistor R2 is connected in series to the collector of T2. Transistor T2 should not conduct as long as the temperature level being monitored is below the set threshold. The base-emitter voltage of T2 decrease by 2 mV per °C. Once this voltage is below the level set by P1, the transistor T2 conducts and the LED begins to light up. By a slow increase of temperature one will notice a dim light from the LED. This way, the user gets an „analog" indication of the temperature.

Power Supply Circuits Sourcebook Vol. 1

The values of R1 and R2 are dependent on the power supply voltage. The values can be computed using the following formulas:

R1 = (Ub/V - 0.6) / 5 kiloohm

R2 = (Ub/V - 1.5) / 15 kiloohm

For example: by 12 volts power supply, the R1 is 2.2 kiloohms while R2 is 680 ohms. The maximum current consumption of the circuit when the LED lights up is 20 mA. Take note that the transistor T2 must not become hotter than 125°C.

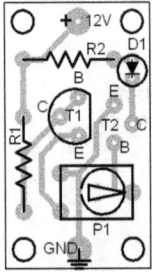

Figure 44.0
Parts Placement Layout

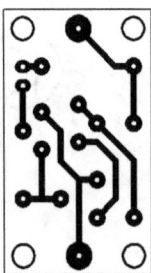

Figure 44.1
Printed Circuit Layout

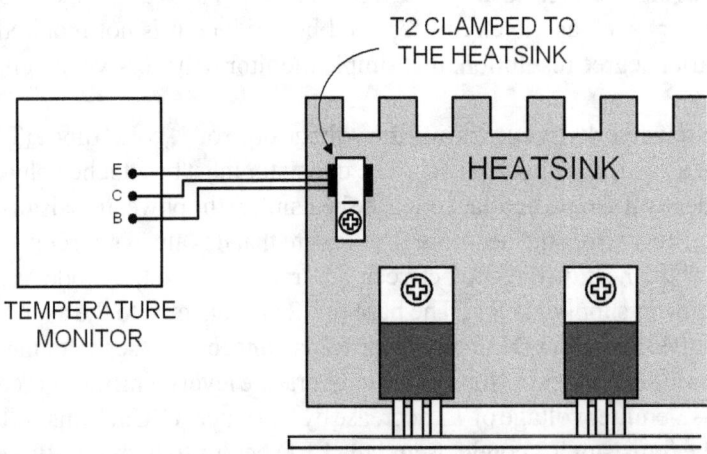

Figure 44.2 External wiring of the T2 temperature sensor and placement to the heatsink being monitored

45 POLARITY INVERTER

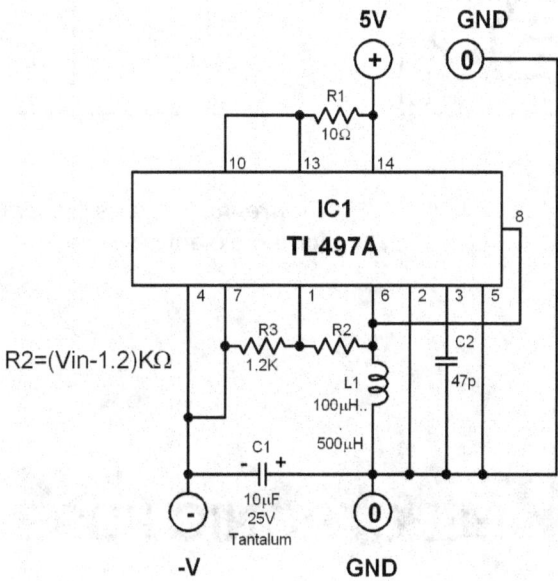

Diagram 45.0 Polarity Inverter

$R2=(Vin-1.2)K\Omega$

This circuit is used to reverse the polarity of a supply line. A negative potential is sometimes necessary in computer or battery powered circuits. Since the current consumption for such a need is mostly low, a single IC can be used as a polarity converter. The TL497A IC is originally designed for downward or upward transformation applications but is used here as polarity converter. The design does not need any transformer or bridge rectifiers.

Coil L1 can be between 100 and 500 μH. The output voltage (Vout) can be found with this formula:

$$Vout = \frac{-1.2}{R2}$$

The load current must not exceed 500 mA.

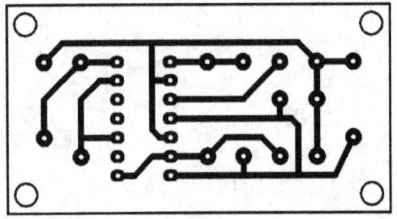

Figure 45.0 Printed Circuit Layout for the Polarity Inverter

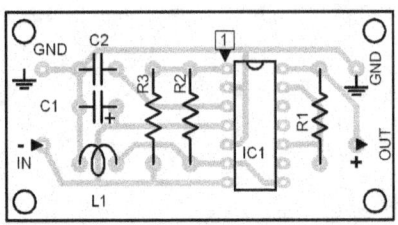

Figure 45.1 Parts Placement Layout for the Polarity Inverter

46 ELECTRONIC FUSE

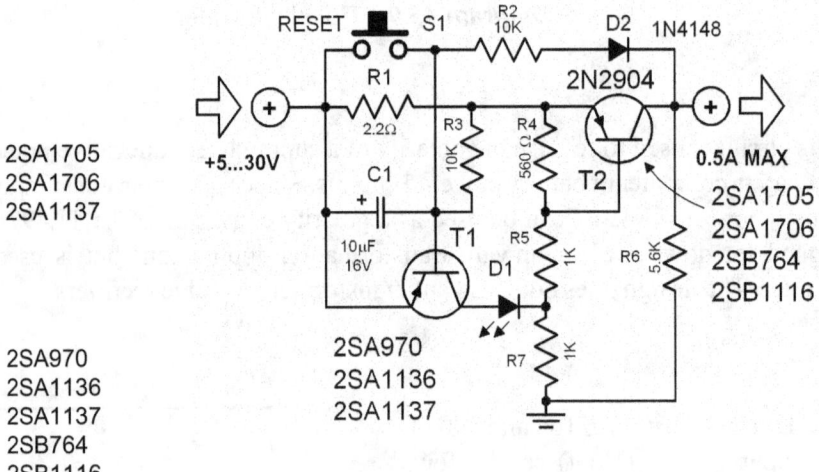

2SA1705
2SA1706
2SA1137

2SA970
2SA1136
2SA1137
2SB764
2SB1116

Diagram 46.0 Electronic Fuse

Of course you know what a fuse is and how it functions. This circuit is the electronic version of a fuse. The electronic fuse cuts off the current (like a normal fuse does) and lights up a LED (not with a normal fuse) when a certain current level is exceeded.

The advantage of this fuse in comparison to a conventional one is that it can be reset and re-used (not to mention the cute lighting of the LED). The circuit functions this way: The current flowing to the connected load flows first through R1. Once the voltage drop at R1 reaches 0.5 volts, transistor T1 conducts and switches off T2 and thereby cutting off the current going out of the circuit. The electronic fuse is at this time "blown" and can only be re-used after pressing S1. You then have enough time to check what is wrong with the device connected to the fuse before resetting it.

The value of R1 in the circuit is selected so that the circuit will "blow" when the current is around 500 mA. The circuit cannot deliver currents above 500 mA due to limitations of the transistors used. If you want to change the limit of the circuit change the value of R1 using this formula: $R1 = 0.4(I)$

Maximum output current is 500 mA.

47 HEATSINK MONITOR

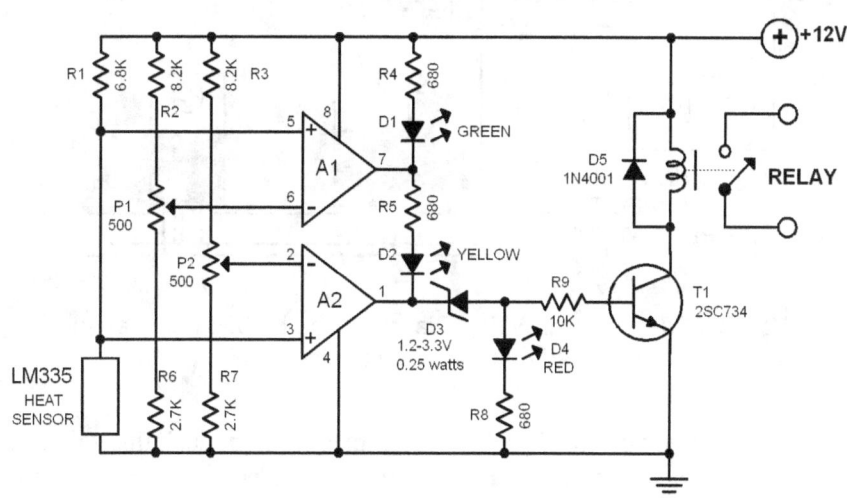

Diagram 47.0 Heatsink Monitor

The heat dissipation from power transistors and diodes in electronic devices are typically absorbed by heatsinks and further dissipated into the environment. The heatsink dimensions are based on the maximum allowable crystal temperature of the semiconductor as published by the manufacturer. However, the actual temperature reached by the heatsink is not clearly known. Many people test the heatsink temperature with their fingers. It seems that as long as it does not burn their fingers, it is accepted to be safe. Obviously, this method is not accurate that is why the electronic circuit featured here was designed.

This heatsink monitor signals via three LEDs when the temperature exceeds two boundary levels. When the heatsink temperature is below 50 ... 60 degrees centigrade (122 ... 140 degrees fahrenheit) the green LED lights. The yellow LED signals that the temperature is within 60 and 70 degrees centigrade (140...158 degrees fahrenheit). The red LED signals that the temperature has reached beyond 70...80 degrees centigrade (158 ...176 degrees fahrenheit). At this high point a relay can be triggered to break the power supply or otherwise protect the electronic device by any other means (e.g. turning on a cooling fan).

As shown in the diagram above the electronic circuit is made of two opamps that are set as window comparators. The input voltage level is set by the thermosensor IC LM335.

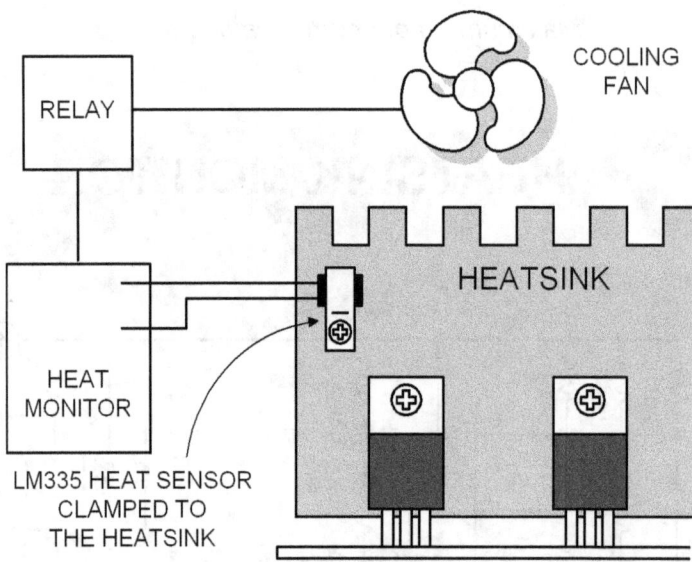

Figure 47.0 External Wiring Diagram. The LM335 heat sensor is clamped to the heatsink being monitored.

This voltage level increases linearly by a factor of 10mV/°centigrade. If this voltage is below the voltage level set by the potentiometers P1 and P2, the opamp outputs are almost at ground potential (0 volts) and the green LED lights up. If this voltage is within the set boundaries, the output of A1 is „high" and the yellow LED will light up. If this voltage exceeds the voltage level set by P2, both A1 and A2 outputs are „high" and the red LED lights up while the yellow LED turns off. At the same time, the load at T1 is reduced to prevent further increase in the heatsink temperature.

To set the desired temperature levels: put the thermosensor LM355 together with a thermometer into a pot of water which is being heated slowly. Turn P1 to its lowest point and turn P2 to its maximum point. The switching point from green LED to yellow LED (50...60°C or 122...140°F) is set with P1. The switching point from yellow LED to red LED is set with P2. Watch the thermometer while the water is being heated and adjust the potentiometers accordingly. Once the desired temperature boundaries are set, attach the thermosensor to the heatsink.

48 COMPUTER CONTROLLED POWER REGULATION

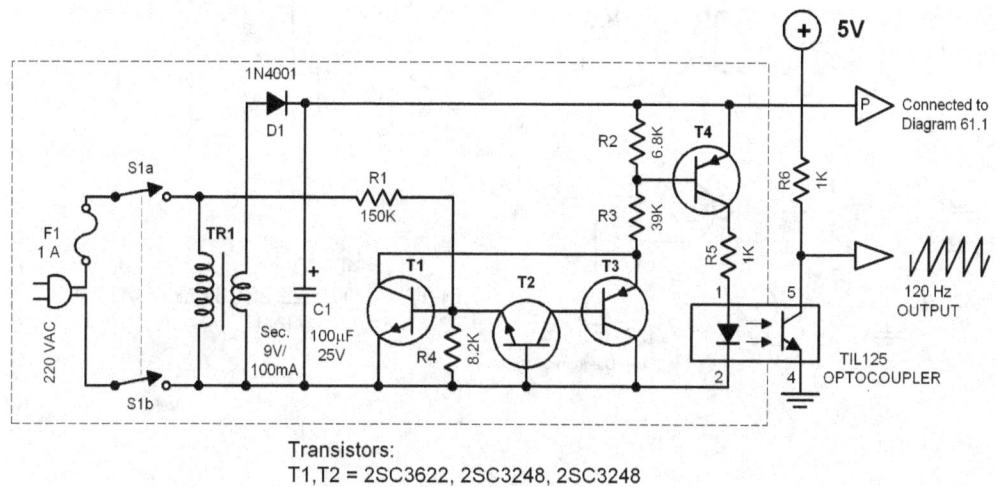

Diagram 48.0 Computer controlled power regulation

This computer interface can control an electrical load like a lamp, electric drill, etc. in 255 steps. The maximum power handling capacity of the circuit is 400 watts. The power regulation is done by controlling the voltage level at the connected load (see Diagram 48.1).

Optocouplers are used to transmit the computer control to the power line part of the circuit. With this technique, the computer is electrically and galvanically isolated from the main power control circuit. This isolation is very important to protect the computer from possible damages caused by high voltage peaks. The phase angle of the voltage that triggers the triac is dependent on the value of the byte delivered by the computer. The higher the byte value, the higher the phase angle becomes, and the shorter will be the time when the triac remains conducting.

Calibration: Turn P1 while the input of the counter sees FFhex (255dec) until the voltage output is 0. The data 00hex produces the same output as the data FFhex. Full power is delivered by data 01hex. The triac must be rated twice the maximum load consumption.

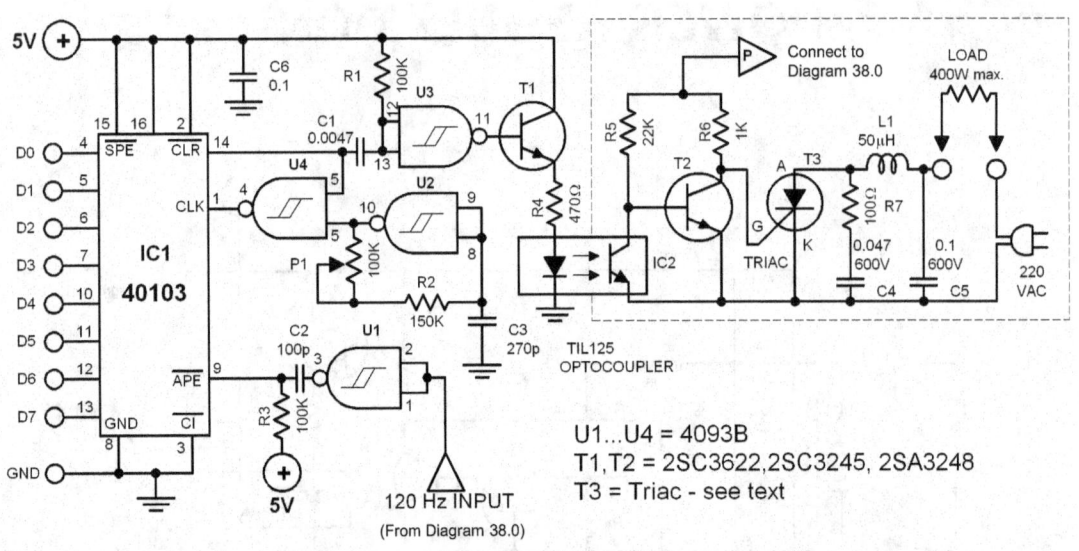

Diagram 48.1 Computer Controlled Power Regulation

2SA 970   2SC3622
2SA1136   2SC3245
2SA1137   2SC3245A
2SC3248

TIC206
TIC216

49   CURRENT ALARM

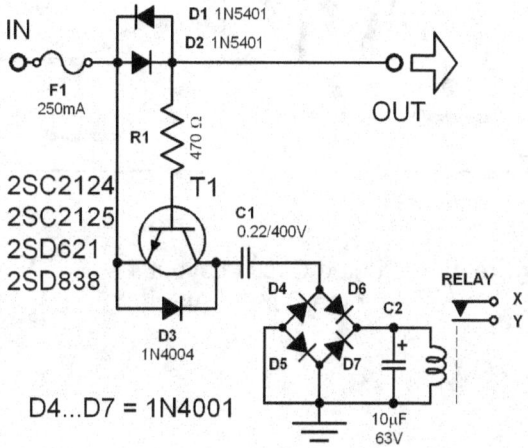

Diagram 49.0   Current Alarm

This electronic circuit is used to monitor the current flow of a device. Once the current flowing through D1 and D2 drops significantly, transistor T1 conducts and the relay closes. The relay then switches on whatever alarm is connected to it.

This circuit has a very wide application area. For example you can monitor exhaust blowers or water pumps. Once the blower or pump motor breaks down, the alarm turns on.

Two PCB layouts are available for this circuit. If you use the first PCB layout, you must install the transistor on a heatsink near the PCB. On the second PCB, you can install the transistor directly on the PCB board. Use a U-shaped heatsink which can be sandwiched between the transistor and the PCB.

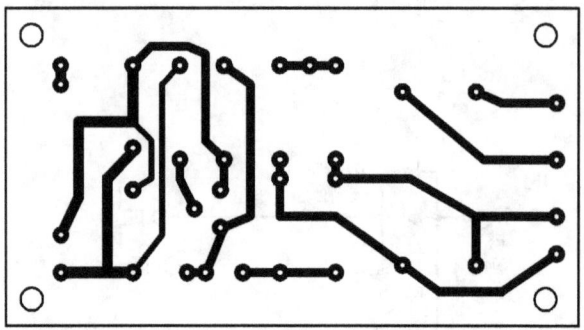

Figure 49.0 Printed Circuit Layout #1
for the Current Alarm

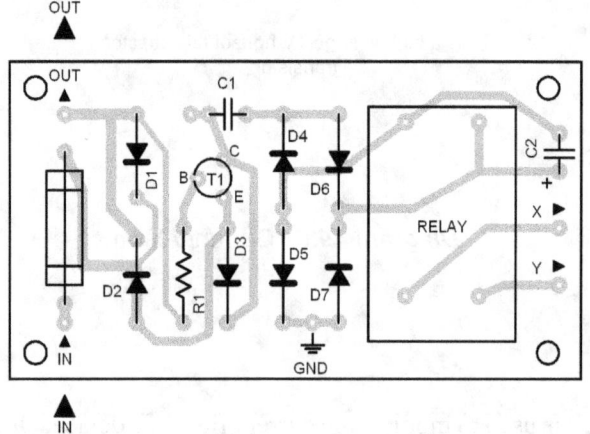

Figure 49.1 Parts Placement Layout #1
for the Current Alarm

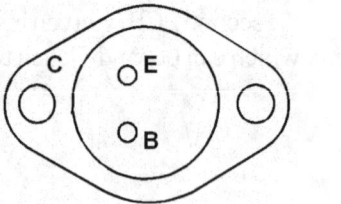

2SC2124
2SC2125
2SD621
2SD838

Note: Bottom view

Parts List:	
Resistors: R1a= 470 Ω **Capacitors:** C1= 0.22/400V ceramic C2= 10µF/63V electrolytic **Diodes:** D1,D2= 1N5401 D3,D4,D5,D6,D7 = 1N4004	**Transistors:** T1= 2SC2124 Replacements: 2SC2125, 2SD621, 2SD838

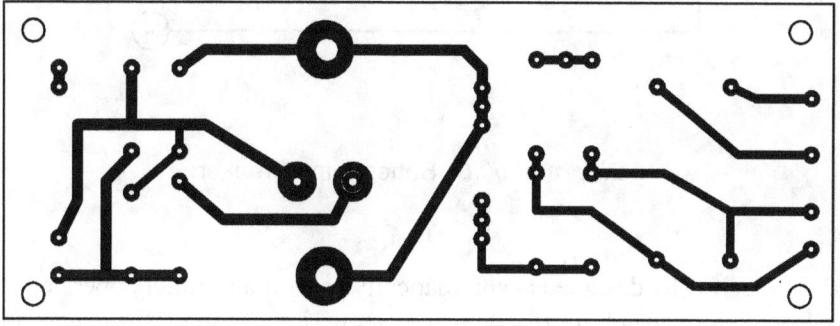

Figure 49.2 Printed Circuit Layout #2 for the Current Alarm

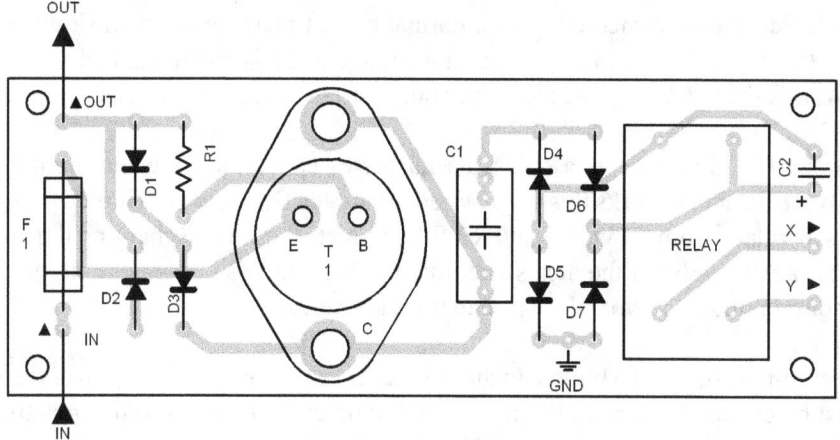

Figure 49.3 Parts Placement Layout #2 for the Current Alarm

50   BATTERY LINE BREAKER

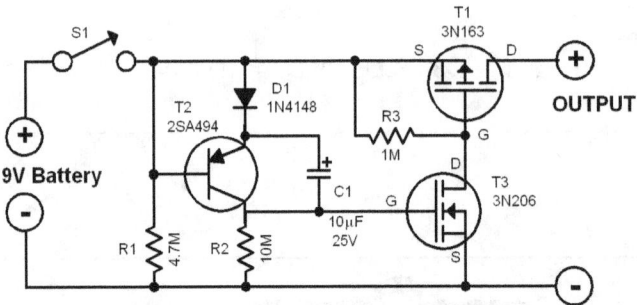

Diagram 50.0 Battery Line Breaker

This circuit is specifically designed to automatically turn off any battery operated device when its user has forgotten to switch it off within a preset period of time. This helps enormously in preserving the battery's life.

How does the electronic circuit work? When the device connected to this circuit is turned on (S1), the capacitor C1 gets a voltage of +9 volts through the diode D1. Since C1 is discharged at the beginning, this voltage goes through to the gate of T3 making both T2 and T3 conduct. The connected device is now turned on as in a normal case. Capacitor C1 then slowly charges via resistor R2. After two or three minutes, the gate voltage of T3 is much reduced to a level that T3 turns off. At this time, T2 turns off the power output to the connected device.

The transistor T1 functions as a quick discharger of capacitor C1 when the user manually switches off the device. It works this way: When the switch S1 is opened (turned off), the base of T1 receives a voltage from C1 via R1 and R2. T1 conducts thereby discharging the capacitor C1. The circuit is again ready for the next switching on. Without this add-on circuit, the user would wait for another minute or two before he can use the device.

Using the circuit is simple. To build it in the device, the positive line of the power (battery) to the device must be cut and the circuit (S1 and output) connected in series to this positive line. The ground line of the circuit must be connected to the minus line of the device.

This circuit can handle current consumptions up to 150 milliamperes. Originally, it was used to control a multimeter device.

This page intentionally left blank

This page intentionally left blank

APPENDICES

Specifications of the transistors used in the projects

Descriptive Part of the Table:

Type
The original type designation has been taken over directly from the manufacturers, with the abbreviation of the manufacturer added in brackets only in those cases in which different manufacturers used the same type designation.

Mat.
The materials used are abbreviated as follows:

Ge	Germanium
Mos	MOS technology (metal oxide silicon)
Si	Silicon
V-MOS	Vertical MOS technology

Pol.
The polarities used are abbreviated as follows:

npn	NPN structure
n-ch	N channel type (FET)
n-p	More than one transistor with different polarities in one case
pnp	PNP structure
p-ch	P channel type (FET)

Abbreviations used in the following table:

A	Antenna amplifer	FET	Field-effect transistor
AGC	Regulating steps	FET-depl.	Field-effecttransistor, depletion type
AF	AF range	FET-enh.	Field-effect transistor, enhancement type
AM	AM range	FM	FM range
CATV	Broad band cable amplifier	fs	Fast switch
CB	CB-radio	HD	Horizontal deflection
CTV	Colour television application	hi-rel	high reliability
chop	Chopper	Idss	Drain source short-circuit current (FET)
Darl	Darlington transistor	IF	IF applications
dg	Dual Gate (FET)	in	Input stages
double	Paired types	iso	insulated
dr	Driver stages	ln	Low noise
dual	Dual transistor (differential amplifier)	min	Miniaturised version
		mix	Mixer stages
end	Final stages	nixie	Digital display tube

osc	Oscillator stages	Ugs	Gate source voltage
pow	Power stages	UHF	UHF range > 250MHz
radiation	Aerospace applications	uni	Universal type
	(radiation-proof)	Up	Pinch-off voltage (FET)
RF	RF range	VD	Vertical deflection
s	Switch	VHF	VHF range 100-250
SMP	Switch-mode power supply	MHz	
SSB	Single side-band operation	Vid	Video output stages
Stabi	Stabilisation	+Diode, +di	With integrated diode
sym	Symmetrical types	../..ns	turn-on/turn-off time
TV	Television applications		

Data Part of the Table:

In the case of the ratings, either average values are quoted (< = max.) or lower (> = min.) guaranteed values. As a rule apply at 25°C, unless otherwise indicated.

Uc

With transistors, the usual situation is for U_{CBO}(colletor base reverse bias) to be quoted, or U_{CEO} and U_{CEO} (collector emitter reverse bias). With FETs, U_{DS} (drain source voltage) is always quoted.

Ic

With transistors, I_c (collector current) is always quoted. If this is followed by (ss) in brackets, I_{CM} is quoted, i.e. the peak value of the collector current. With FETs, I_D (drain current) is always quoted.

Ptot

As a rule, the total leakage power Ptot is quoted, with RF types we always quote the RF output power P_Q, with corresponding frequency in brackets.

Amplification

The DC current gain B(h_{FE}) or the short-circuit current gain ß(h_{fe}) are always quoted as guaranteed values.

fт

The transition frequency is always qouted in MHz.

Specifications of the transistors used in the projects

Type	Mat.	Pol.	Description	UC [Vmax]	IC [Amax]	Ptot [Wmax]	Current Gain	fT [MHz]
MJ3001	Si	npn	Darl+diode,pow	60	10.00	150.00($25°C)	>10	
MJE243	Si	npn	AF-s-pow	100	4.00	1.50($25°C)	40.120	>40.00
MJE244	Si	npn	AF-s-pow	100	4.00	1.50($25°C)	>25	>40.00
MJE253	Si	npn	AF-s-pow	100	4.00	1.50($25°C)	40-120	>40.00
MJE4350	Si	pnp	AF-end,s-pow	100	16.00	125.00($25°C)	15	>1.00
MJE5170	Si	pnp	uni-pow	120	6.00	2.00($25°C)	15-100	>1.00
MJE5180	Si	npn	uni-pow	120	6.00	2.00($25°C)	15-100	>1.00
MPF102	Si	n-ch	FET,VHF-in,sym,mix 25V, Idss>2mA,Up<V					
MPF106	Si	n-ch	FET,VHF 25V,Idss>4mA,Up<8V					
MPS-A29	Si	npn	Darl	100	0.50	1.50($25°C)	>10	>125.00
2N708	Si	npn	s	40/15	0.20	1.20(25°C)	>15	480.00
2N1711	Si	npn	uni	75	0.50	3.00(25°C)	75	>70.00
2N1889	Si	npn	AF-s	100/60	0.50	3.00(25°C)	40-120	>50.00
2N1890	Si	npn	AF-s	100/60	0.50	3.00(25°C)	100-300	>60.00
2N1990	Si	npn	nixie	100	1.00	2.00(25°C)	>25	
2N2102	Si	npn	AF-s	120/65	1.00	5.00(25°C)	40-120	>120.00
2N2222	Si	npn	ini	0		1.80(25°C)		
2N2368	Si	npn	fs	40/15	0.20	1.20(25°C)	20-60	>400.00
2N2369	Si	npn	fs	40/15	0.20	1.20(25°C)	40-120	>500.00
2N2905	Si	pnp	uni	60/40	0.60	3.00(25°C)	100-300	>200.00
2N2904	Si	pnp	uni	60/40	0.60	3.00(25°C)	40-120	>200.00
2N3019	Si	npn	uni	140/80	1.00	5.00(25°C)	100-300	>100.00
2N3020	Si	npn	uni	140/80	1.00	5.00(25°C)	40-120	>80.00
2N3055	Si	npn	AF-s-pow	100/60	15.00	115.00($25°C)	20-70	>2.50
2N3109	Si	npn	AF-s	80/40	1.00	5.00(25°C)	100-300	>70.00
2N3110	Si	npn	AF-s	80/40	1.00	5.00(25°C)	40-120	>60.00
2N3367	Si	n-ch	FET,uni,In	40V,Idss>0.5mA,Up<2.5V				
2N3370	Si	n-ch	FET,uni,In	40V,Idss>0.1mA,Up3.2V				
2N3454	Si	n-ch	FET,uni	50V,Idss>0.05mA,Up<2.3V				
2N3819	Si	n-ch	FET,VHF,uni,sym	25V,Idss>2mA,Up<8V				
2N3823	Si	n-ch	FET,VHF,In	30V,Idss>4mA,Up<8V				
2N3903	Si	npn	uni	60/40	0.20	1.50(25°C)	50-150	>250.00
2N3904	Si	npn	uni	60/40	0.20	1.50(25°C)	100-300	>300.00
2N3905	Si	pnp	uni	40	0.20	1.50(25°C)	50-150	>200.0
2N3906	Si	pnp	uni	40	0.20	1.50(25°C)	100-300	>250.00
2N4118	Si	n-ch	FET,uni	40V,Idss>0.08mA,Up<3V				
2N5294	Si	npn	AF-s-pow	80/70	4.00	1.80(25°C)	30-120	>0.80
2N5397	Si	n-ch	FET,VHF/UHF	25V,Idss>10mA,Up<6V				
2N5398	Si	n-ch	FET,VHF/UHF	25V,Idss>5mA,Up<6V				
2N5486	Si	n-ch	FET,VHF/UHF	25V,Idss>8mA,Up<6V				
2N6038	Si	npn	Darl+diode,pow	60	4.00	1.50($25°C)	>10	>25.00
2N6039	Si	npn	Darl+diode,pow	80	4.00	1.50($25°C)	>10	>25.00
2N6283	Si	npn	Darl+diode,pow	80	20.00	160.00($25°C)	>10	>4.00
2N6284	Si	npn	Darl+diode,pow	100	20.00	160.0($25°C)	>10	>4.00
2N6412	Si	npn	AF-s-pow	60/40	4.00	15.00($25°C)	>5	>50.00
2N6414	Si	pnp	AF-s-pow	80/60	4.00	15.00($25°C)	>5	>50.00
2SA511	Si	pnp	AF/RF/s	90/80	1.50	8.00(25°)	30-150	60.00
2SA597	Si	pnp	RF-s	50/40	1.00	6.00($25°C)	10-250	400.00
2SA761	Si	pnp	uni	110	2.00	6.30($25°)	50-240	80.00
2SA970	Si	pnp	AF,In	120	0.10	0.30(25°C)	200-700	100.0

Specifications of the transistors used in the projects

Type	Mat.	Pol.	Description	UC [Vmax]	IC [Amax]	Ptot [Wmax]	Current Gain	fT [MHz]
2SA1016	Si	pnp	uni,ln	120/100	0.05	0.40(25°)	160-960	110.00
2SA1123	Si	pnp	uni,ln	150	0.05	0.7(25°)	65-450	200.00
2SA1136	Si	pnp	AF-in,ln	120/100	0.10	0.30(25°C)	120-560	90.00
2SA1137	Si	pnp	AF-in,on	80	0.10	0.30(25°C)	120-560	90.00
2SA1141	Si	pnp	AF/Rf-pow	115	10.00	2.00($25°C)	100	80.00
2SA1285	Si	pnp	uni	120	0.20	0.90(25°C)	150-800	200.00
2SA1285A	Si	pnp	uni	150	0.10	0.90(25°C)	150-500	200.00
2SA1515	Si	pnp	uni	40/32	1.00	0.50(25°C)	82-390	150.00
2SA1705	Si	pnp	AF,s	60/50	1.00	0.90(25°C)	>30	150.00
2SA1706	Si	pnp	AF-s	60/50	2.00	1.00(25°C)	>40	150.00
2SB633	Si	pnp	AF-s-pow	100/85	6.00	40.00($25°C)	40-320	15.00
2SB764	Si	pnp	uni	60/50	1.00	0.90(25°C)	60-320	150.00
2SB822	Si	pnp	Af-dr/end	40/32	2.00	0.75(25°C)	82-390	100.00
2SB826	Si	pnp	s-pow	60/50	7.00	60.00($25°C)	>30	10.00
2SB867	Si	pnp	AF/s-pow,lo-sat	130/80	3.00	30.00($25°C)	60-260	30.00
2SB868	Si	pnp	AF/s-pow,lo-sat	130/80	4.00	35.00($25°C)	60-260	30.00
2SB869	Si	pnp	AF/s-pow,lo-sat	130/80	5.00	40.00($25°C)	60-260	30.00
2SB870	Si	pnp	AF/s-pow,lo-sat	120/80	7.00	40.00($25°C)	60-260	30.00
2SB874	Si	pnp	AF/s-pow, TV-VD	100/60	2.00	20.00($25°C)	>40	250.00
2SB909	Si	pnp	AF-dr/end	40/32	1.00	1.00(25°C)	82-390	150.00
2SB911	Si	pnp	AF-dr/end	40/32	2.00	1.00(25°C)	82-390	100.0
2SB920	Si	pnp		120/80				
2SB921	Si	pnp		120/80				
2SB1064	Si	pnp	AF-s-pow	60/50	3.00	1.50($25°)	60-320	70.00
2SB1114	Si	pnp	min,uni	20	2.00	2.00($25°C)	135-600	180.00
2SB1116	Si	pnp	uni	60/50	1.00	0.75(25°C)	135-600	120.00
2SB1142	Si	pnp	s-pow	60/50	2.50	10.00(25°C)	>35	140.00
2SB1143	Si	pnp	s-pow	60/50	4.00	10.00(25°C)	>40	150.00
2SB1144	Si	pnp	AF/s-pow,lo-sat	120/100	1.50	10.00(25°C)	>30	100.00
2SB1230	Si	pnp	AF/s-pow,lo-sat	110/100	15.00	100.00($25°C)	50-140	
2SB1231	Si	pnp	AF/s-pow,lo-sat	110/100	25.00	120.00($25°C)	50-140	
2SB1232	Si	pnp	AF/s-pow,lo-sat	110/100	40.00	150.00($25°C)	50-140	
2SC270	Si	npn	s-pow	270/75	5.00	50.00($25°C)	24-40	22.00
2SC460	Si	npn	AM-in/mix/osc	30	0.10	0.20(25°C)	35-200	230.00
2SC696	Si	npn	uni	100/60	3.00	0.75(25°C)	30-173	100.00
2SC763	Si	npn	VHF	25/12	0.02	0.10(25°C)	20-300	>400.00
2SC829	Si	npn	AM/FM-in/mix/osc	30/20	0.03	0.40(25°C)	40-500	230.00
2SC959	Si	npn	uni	120/80	0.70	0.70(25°C)	40-200	100.00
2SC1324	Si	npn	UHF-CATV	35/25	0.15	3.00(25°C)	10-35	
2SC1876	Si	npn	Darl	100/70	0.50	0.80(25°C)	>20	
2SC2124	Si	npn	TV-HD	220/800	2.00	5.00($90°C)	20	4.00
2SC2125	Si	npn	TV-HD	220/800	5.00	50.00($25°C)	8-25	5.00
2SC2270	Si	npn	lo-sat	50/20	5.00	1.00($25°C)	>70	100.00
2SC2334	Si	npn	s-pow,dc-dc conv.	150/100	7.00	40.00($25°C)	>20	
2SC2459	Si	npn	uni	120	0.10	0.20(25°C)	200-700	100.00
2SC2675	Si	npn	AF,ln	80	0.10	0.30(25°C)	180-820	120.00
2SC2724	Si	npn	FM-IF	30/25	0.03	0.20(25°C)	25-300	200.00
2SC3112	Si	npn	AF,ln	50	0.15	0.40(25°C)	600-3600	250.00
2SC3179	Si	npn	AF-pow	80/60	4.00	30.00($25°C)	100	15.00
2SC3245	Si	npn	uni	120	0.10	0.90(25°C)	150-800	200.00

Specifications of the transistors used in the projects

Type	Mat.	Pol.	Description	UC [Vmax]	IC [Amax]	Ptot [Wmax]	Current Gain	fT [MHz]
2SC3245A	Si	npn	uni	150	0.10	0.90(25°C)	400-800	200.00
2SC3248	Si	npn	uni	180	0.10	0.90(25°C)	150	130.00
2SC3358	Si	npn	UHF	20/12	0.10	0.25(25°C)	50-300	7000.00
2SC3420	Si	npn	lo-sat	50/20	5.00	10.00(25°C)	>70	100.00
2SC3622	Si	npn	AF-s,hi-beta	60/50	0.15	0.25(25°C)	1000-3200	250.00
2SC4308	Si	npn	VHF-A	30/20	0.30	0.60(25°C)	50-200	2500.00
2SD386	Si	npn	TV-VD	200/120	3.00	1.75($25°C)	40-320	8.00
2SD406	Si	npn	Darl	100	2.00	15.00(25°C)	>2000	
2SD613	Si	npn	AF-s-pow	100/85	6.00	40.00($25°C)	40-320	15.00
2SD614	Si	npn	Darl	100/80	3.00	0.80(25°C)	3000	15.00
2SD621	Si	npn	TV_HD	2500/900	3.00	50.00($25°C)	3-15	
2SD628	Si	npn	Darl+diode,pow	100	10.00	80.00($25°C)	>1000	
2SD629	Si	npn	Darl+diode,pow	100	10.00	100.00($25°C)	>1000	
2SD688	Si	npn	Darl,pow	100	1.50	0.80($25°C)	>10	
2SD712	Si	npn	AF-s-pow	100	4.00	30.00($25°C)	55-300	8.00
2SD726	Si	npn	AF-s-pow	100/80	4.00	40.00($25°C)	35-320	10.00
2SD729	Si	npn	Darl+diode,pow	100	20.00	125.00($25°C)	>1000	
2SD781	Si	npn	s-pow,TV-HD	150/60	2.00	1.00(25°C)	150	
2SD826	Si	npn		60/20	5.00	1.00($25°C)	120-560	120.00
2SD838	Si	npn	TV-HD,s-pow	2500/900	3.00	50.00($25°C)	3-15	
2SD892A	Si	npn	Darl	60/50	0.50	0.40(25°C)	>8000	150.00
2SD1049	Si	npn	AF-s-pow	120/80	25.00	80.00($25°C)	>20	
2SD1062	Si	npn	s-pow	60/50	12.00	40.00($25°C)	>30	10.00
2SD1153	Si	npn	Darl	80750	1.50	0.90(25°C)	>40	120.00
2SD1177	Si	npn	AF-pow,TV-HD	100/60	2.00	20.00($25°C)	>40	230.00
2SD1237	Si	npn	s-pow	90/80	7.00	1.75($25°C)	>30	20.00
2SD1238	Si	npn	s-pow	90/80	12.00	80.00($25°C)	>30	20.00
2SD1639	Si	npn	AF-s-pow	100/80	2.20	10.00($25°C)	40-200	
2SD1684	Si	npn	AF/s-pow,lo-sat	120/100	1.50	10.00(25°C)	>30	120.00
2SD1685	Si	npn	AF/s-pow,lo-sat	60/20	5.00	10.00(25°C)	>95	120.00
2SD1691	Si	npn	AF-s-pow	60	5.0	20.00(25°C)	100-400	
2SD1840	Si	npn	AF/s-pow,lo-sat	110/100	15.00	100.00($25°C)	50-140	
2SD1841	Si	npn	AF/s-pow,lo-sat	110/100	25.00	120.00($25°C)	50-140	
2SD1842	Si	npn	AF/s-pow,lo-sat	110/100	40.00	150.00($25°C)	50-140	
2SD2116	Si	npn	Darl	80/50	0.70	1.00(25°C)	>40	
2SD2117	Si	npn	Darl	80/50	1.50	1.00(25°C)	>30	
2SD2213	Si	npn	Darl,AF	150/80	1.50	0.90(25°C)	>10	
2SJ165	V-MOS	p-ch	FET-enh.,	50V,0.1A,0.25W				
2SK422	V-MOS	n-ch	FET-enh.	60v,0.7A,0.9W,17/12ns				
2SK423	V_MOS	n-ch	FET-enh.	100V,0.5A,0.9W,15/20ns				
3N140	MOS	n-ch	FET-depl.,dg,FM/VHF-in	20V,Idss>5mA				
3N225	MOS	n-ch	FET-depl.,dg, UHF	25V,Idss>1mA,Up<4V				
3SK35	MOS	n-ch	FET-depl.,dg,VHF	20V,Idss>3mA,Up<4V				
3SK37	MOS	n-ch	FET-depl.,dg,VHF	20V,Idss>4mA,Up<3V				
3SK45	MOS	n-ch	FET-depl.,dg,VHF	22V,Idss>4mA,Up<3V				
3SK61	MOS	n-ch	FET-depl.,dg,VHF	20V,Idss>4mA,Up<3V				
3SK72	MOS	n-ch	FET-depl.,dg,VHF	20V,Idss>2.5mA,Up<3V				
3SK77	MOS	n-ch	FET-depl.,dg,VHF	20V,Idss>3mA,Up<2.5V				
3SK85	MOS	n-ch	FET-depl.,dg,VHF	20V,Idss>4mA,Up<3V				

SEMICONDUCTOR DIODE SPECIFICATIONS

			Peak Inverse Voltage, PIV (Volts)	Average Rectified Current Forward (Reverse) IO (A) (IR(A))	Peak Surge Current, IFSM 1 sec. @ 25°C (A)	Average Forward Voltage, VF (Volts)
* RFR = Rectifier, Fast Recovery						
Device	**Type**	**Material**				
1N34	Signal	Germanium	60	8.5 m (15.0µ)		1.0
1N34A	Signal	Germanium	60	5.0 m (30.0µ)		1.0
1N67A	Signal	Germanium	100	4.0 m (5.0µ)		1.0
1N191	Signal	Germanium	90	5.0 m	1.0	
1N270	Signal	Germanium	80	0.2 (100 µ)		1.0
1N914	Fast Switch	Silicon (Si)	75	75.0 m (25.0 n)	0.5	1.0
1N1184	RFR	Si	100	35 (10 m)		1.7
1N2071	RFR	Si	600	0.75 (10.0µ)		0.6
1N3666	Signal	Germanium	80	0.2 (25.0µ)		1.0
1N4001	RFR	Si	50	1.0 (0.03 m)		1.1
1N4002	RFR	Si	100	1.0 (0.03 m)		
1N4003	RFR	Si	200	1.0 (0.03 m)		1.1
1N4004	RFR	Si	400	1.0 [0.03 m)		1.1
1N4005	RFR	Si	600	1.0 (0.03 m)		1.1
1N4006	RFR	Si	800	1.0 (0.03 m)		1.1
1N4007	RFR	Si	1000	1.0 (0.03 m)		1.1
1N4148	Signal	Si	75	10.0 m (25.0 n)		1.0
1N4149	Signal	Si	75	10.0 m (25.0 n)		1.0
1N4152	Fast Switch	Si	40	20.0 m (0.05µ)		0.8
1N4445	Signal	Si	100	0.1 (50.0 n)		1.0
1N5400	RFR	Si	50	3.0	200	
1N5401	RFR	Si	100	3.0	200	
1N5402	RFR	Si	200	3.0	200	
1N5403	RFR	Si	300	3.0	200	
1N5404	RFR	Si	400	3.0	200	
1N5405	RFR	Si	500	3.0	200	
1N5406	RFR	Si	600	3.0	200	
1N5767	Signal	Si		0.1 (1.0µ)		1.0
ECG5863		RFR	Si	600 6	150	0.9

ZENER DIODES SPECIFICATIONS

Zener Voltage (Volts)	Power (Watts)							
	0.25	0.4	0.5	1.0	1.5	5.0	10.0	50.0
1.8	1N4614							
2.0	1N4615							
2.2	1N4616							
2.4	1N4617	1N4370,A	1N4370,A,1N5221,B 1N5985,B					
2.5			1N5222B					
2.6	1N702,A							
2.7	1N4618	1N4371,A	1N4371,A,1N5223,B 1N5839, 1N5986					
2.8			1N5224B					
3.0	1N4619	1N4372,A	1N4372,1N5225,B 1N5987					
3.3	1N4620	1N746,A 1N764 A 1N5518	1N746A 1N5226,B 1N5988	1N3821 1N4728,A	1N5913	1N5333,B		
3.6	1N4621	1N747,A 1N5519	1N747A 1N5227,B,1N5989	1N3822 1N4729,A	1N5914	1N5334,B		
3.9	1N4622	1N748,A 1N5520	1N748A,1N5228,B 1N5844, 1N5990	1N3823 1N4730,A	1N5915	1N5335,B	1N3993A	1N4549,B 1N4557,B
4.1	1N704,A							
4.3	1N4623	1N749,A 1N5521	1N749,A 1N5229,B 1N5845,1N5991	1N3824 1N4731,A	1N5916	1N5336,B	1N3994,A	1N4550,B 1N4558,B
4.7	1N4624	1N750,A 1N5522	1N750A,1N5230,B 1N5846, 1N5992	1N3825 1N4732,A	1N5917	1N5337,B	1N3995,A	1N4551,B 1N4559,B
5.1	1N4625 1N4689	1N751 A 1N5523	1N751A, 1N5231,B 1N5847,1N5993	1N3826 1N4733	1N5918	1N5338,B 1N4560,B	1N3996,A	1N4552,B
5.6	1N708A 1N4626	1N752,A 1N5524	1N752,A,1N5232,B 1N5848, 1N5994	1N3827 1N4734,A	1N5919	1N5339,B 1N4561,B	1N3997,A	1N4553,B
5.8	1N706A	1N762						
6.0				1N5233B 1N5849			1N5340,B	
6.2	1N709,1N4627 MZ605,MZ610 MZ620,MZ640	1N753,A 1N821,3,5, 7,9; A	1N753,A 1N5234,B, 1N5850 1N5995	1N3828,A 1N4735,A	1N5920	1N5341,B 1N4562,B	1N3998,A	1N4554,B
6.4	1N4565-84,A							
6.8	1N4099	1N754,A 1N957,B 1N5526	1N754,A 1N757,B 1N5235,B 1N5851 1N5996	1N3016,B 1N3829 1N4736,A	1N3785 1N5921	1N5342,B	1N2970,B 1N3999,A	1N2804B 1N3305B 1N4555, 1N4563
7.5	1N4100	1N755,A 1N958,B 1N5527	1N755A,1N958,B 1N5236,B, 1N5862 1N5997	1N3017,A,B 1N3830 1N4737,A	1N3786 1N5922	1N5343,B 1N4000,A 1N4556,	1N2971,B 1N3306,B	1N2805,B 1N4564
8.0	1N707A							
8.2	1N712A 1N4101	1N756,A 1N959,B 1N5528	1N756,A 1N959,B,1N5237,B 1N5853,1N5998	1N3018,B 1N4738,A	1N3787 1N5923	1N5344,B	1N2972,B	1N2806,B 1N3307,B
8.4		1N3154-57,A 1N3155-57	1N3154,A					
8.5	1N4775-84,A		1N5238,B,1N5854					
8.7	1N4102					1N5345,B		
8.8		1N 764						
9.0		1N764A	1N935-9;A,B					

ZENER DIODES SPECIFICATIONS

Zener Voltage (Volts)	Power (Watts)							
	0.25	0.4	0.5	1.0	1.5	5.0	10.0	50.0
9.1	1N4103	1N757,A 1N960,B 1N5529	1N757,A, 1N960,B 1N5239,B, 1N5855 1N5999	1N3019,B 1N4739,A	1N3788 1N5924	1N5346,B	1N2973,B	1N2807,B 1N3308,B
10.0	1N4104	1N758,A 1N961,B 1N5530,B	1N758,A, 1N961,B 1N5240,B, 1N5856 1N6000	1N3020,B 1N4740	1N3789 1N5925	1N5347,B	1N2974,B	1N2808,B 1N3309,A,B
11.0	1N715,A 1N4105	1N962,B 1N5531	1N962,B,1N5241,B 1N5857, 1N6001	1N3021,B 1N4741,A	1N3790 1N5926	1N5348,B	1N2975,B	1N2809,B 1N3310,B
11.7	1N716,A 1N4106		1N941,A,B					
12.0		1N759,A 1N963,B 1N5532	1N759,A,1N963,B 1N5242,B, 1N5858 1N6002	1N3022,B 1N4742,A	1N3791 1N5927	1N5349,B	1N2976,B	1N2810,B 1N3311,B
13.0	1N4107	1N964,B 1N5533	1N964,B,1N5243,B 1N5859,1N6003	1N3023,B 1N4743,A	1N3792 1N5928	1N5350,B	1N2977,B	1N2811,B 1N3312,B
14.0	1N4108	1N5534	1N5244B, 1N5860			1N5351,B	1N2978,B	1N2812,B 1N3313,B
15.0	1N4109	1N965,B 1N5535	1N965,B,1N5245,B 1N5861,1N6004	1N3024,B 1N4744A	1N3793 1N5929	1N5352,B	1N2979,A,B	1N2813,A,B 1N3314,B
16.0	1N4110	1N966,B 1N553,B	1N966,B,1N5246,B 1N5862, 1N6005	1N3025,B 1N4745,A	1N3794 1N5930	1N5353,B	1N2980,B	1N2814,B 1N3315,B
17.0	1N4111	1N5537	1N5247,B 1N5863			1N5354,B	1N2981B	1N2815,B 1N3316,B
18.0	1N4112	1N967,B 1N5538	1N967,B 1N5248,B 1N5864, 1N6006	1N3026,B 1N4746,A	1N3795 1N5931	1N5355,B	1N2982,B	1N2816,B 1N3917,B
19.0	1N4113	1N5539	1N5249,B 1N5865			1N5356,B	1N2983,B	1N2817,B 1N3318,B
20.0	1N4114	1N968,B 1N5540	1N968,B,1N5250,B 1N5866, 1N6007	1N3027,B 1N4747,A	1N3796 1N5932,A,B	1N5357,B	1N2984,B	1N2818,B 1N3319,B
22.0	1N4115	1N959,B 1N5541	1N969,B,1N5241,B 1N5867, 1N6008	1N3028,B 1N4748,A	1N3797 1N5933	1N5358,B	1N2985,B	1N2819,B 1N3320,A,B
24.0	1N4116	1N5542 1N9701B	1N970,B,1N5252,B 1N586,1N6009	1N3029,B 1N4749,A	1N3798 1N5934	1N5359,B	1N2986,B	1N2820,B 1N3321,B
25.0	1N4117	1N5543	1N5253,B 1N5869			1N5360,B	1N2987B	1N2821,B 1N3322,B
27.0	1N4118	1N971,B	1N971,1N5254,B 1N5870,1N6010	1N3030,B 1N4750,A	1N3799 1N5935	1N5361,B	1N2988,B	1N2822B 1N3323,B
28.0	1N4119	1N5544	1N5255,B,1N5871			1N5362,B		
30.0	1N4120	1N972,B 1N5546	1N972,B,1N5256,B 1N5872,1N6011	1N3031,B 1N4751,A	1N3800 1N5936	1N5363,B	1N2989,B	1N2823,B 1N3324,B
33.0	1N4121	1N973,B 1N5546	1N973,B,1N5257,B 1N5873,1N6012	1N3032,B 1N4752,A	1N3801 1N5937	1N5364,B	1N2990,A,B	1N2824,B 1N3325,B
36.0	1N4122	1N974,B	1N974,B,1N5258,B 1N5874,1N6013	1N3033,B 1N4753,A	1N3802 1N5938	1N5365,B	1N2991,B	1N2825,B 1N3326,B
39.0	1N4123	1N975,B	1N975,B, 1N5259,B 1N5875 ,1N6014	1N3034,B 1N4754,A	1N3803 1N5939	1N5366,B	1N2992,B	1N2826,B 1N3327,B
43.0	1N4124	1N976,B	1N976,B,1N5260,B 1N5876,1N6015	1N3035,B 1N4755,A	1N3804 1N5940	1N5367,B	1N2993,A,B	1N2827,B 1N3328,B
45.0			1N2994B	1N2828B 1N3329B				

POWER FETs

Device No.	Type	Max. Diss. (W)	Max. V_{DS} (Volts)	Max. I_D (A)*	Gfs mmhos (typ.)	Input C_{iss} (pF)	Output C_{oss} (pF)	Approx. Upper Freq. (MHz)	Case	Pack-Type Mnfr.	General applications age/
DV1202S	N-Chan.	10	50	0.5	100k	14	20	500	.380 SOE	1/S	RF power amp., oscillator
DV1202W	N-Chan.	10	50	0.5	100k	14	20	500	C-220	5/S	RF power amp., oscillator
DV1205S	N-Chan.	20	50	1	200k	26	38	500	.380 SOE	1/S	RF power amp., oscillator
DV1205W	N-Chan.	20	50	1	200k	26	98	500	C-220	5/S	RF power amp., oscillator
2SK133	N-Chan.	100	120	7	1M	600	350	1	TO-3	6/H	AF pwr. amp., switch (complem to 2SJ48)
2SK134	N-Chan.	100	140	7	1M	600	350	1	TO-3	6/H	AF pwr. amp., switch (complem to 2SJ49)
2SK135	N-Chan.	100	160	7	1M	600	350	1	TO-3	6/H	AF pwr. amp., switch (complem to 2SJ50)
2SJ48	P-Chan.	100	120	7	1M	900	400	1	TO-3	6/H	AF pwr. amp., switch (complem to 2SK133)
2SJ49	P-Chan.	100	140	7	1M	900	400	1	TO-3	6/H	AF pwr. amp., switch (complem to 2SK134)
2SJ50	P-chan.	100	160	7	1M	900	400	1	TO-3	6/H	AF pwr. amp., switch (complem to 2SK135)
VMP4	N-Chan.	25	60	2	170K	32	4.8	200	.380 SOE	1/S	VHF pwr. amp., rcvr front end (rf amp., mixer).
VN10KM	N-Chan.	1	60	0.5	100K	48	16	–	TO-92	2/S	High-speed line driver, relay driver, LED stroke driver
VN64GA	N-Chan.	80	60	12.5	150K	700	325	30	TO-3	3/S	Linear amp., power-supply switch, motor control
VN66AF	N-Chan.	15	60	2	150K	50	50	–	TO-202	4/S	High-speed switch, HF linear amp., audio amp. line driver.
VN66AK	N-Chan.	8.3	60	2	250K	93	6	100	TO-39	7/S	RF pwr.amp.,high-current analog switching
VN67AJ	N-Chan.	25	60	2	250K	33	7	100	TO-3	3/S	RF pwr.amp.,high-current switching
VN89AA	N-Chan.	25	80	2	250K	50	10	100	TO-3	3/S	High-speed switching,HF linear amps., line drivers.
IRF100	N-Chan.	125	80	16	300K	900	25	–	TO-3	3/S	High-speed switching,audio inverters.
IRF101	N-Chan.	125	60	16	300K	900	25	–	TO-3	3/S	Same as IRF100

Legend: * 25°C (case) S = M/A-COM IR = International Rectifier. H = Hitachi Mnfr = Manufacturer

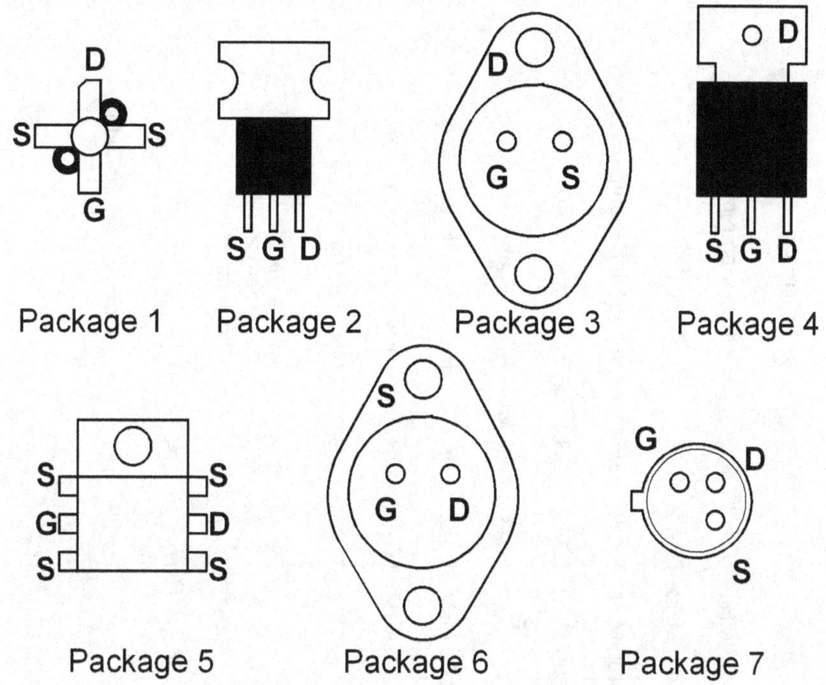

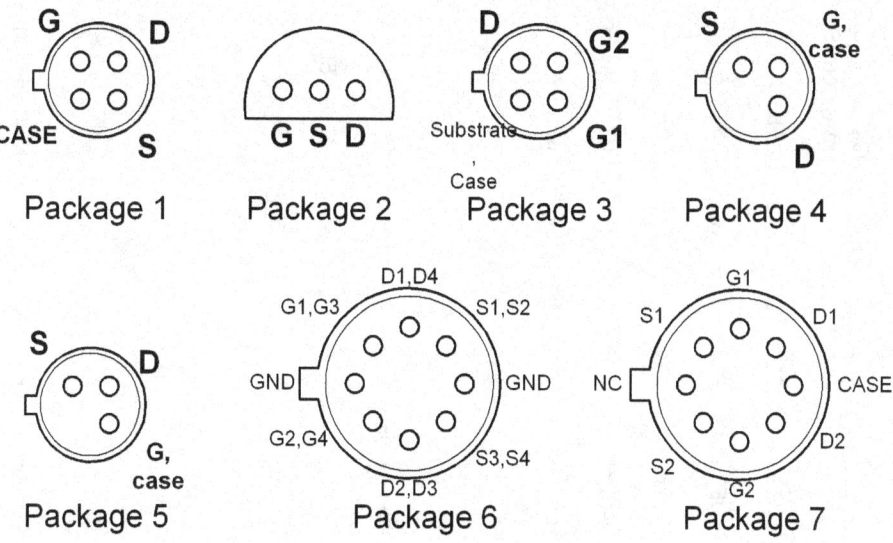

Package Information for Power FETs

Package Information for Small Signal FETs

SMALL-SIGNAL FETs

Device No.	Type	Max. Diss. (mW)	Max. V_{DS} (Volts)	Max. I_D	Min G_{fs} (mA)* (mS)	Input C (pF)	$V_{GS(off)}$	Upper Freq. (volts)(MHz)	Noise Figure (MHz)	Case Type (typ)	/Mnfr.	General applications
2N4416	N-JFET	300	30	-15	4.5K	4	-6	450	400 MHz 4 dB	TO-72	1/S,M	VHF/UHF/RF amp.mix., osc.
2N5484	N-JFET	310	25	30	2.5K	5	-3	200	200 MHz 4 dB	TO-92	2/M	VHF/UHFamp,mix., osc.
2N5485	N-JFET	310	25	30	3.5K	5	-4	400	400 MHz 4 dB	TO-92	2/S	VHF/UHF/RF amp.mix., osc.
3N200	N-Dual-Gate MOSFET	330	20	50	10K	4-8.5	-6	500	400 MHz 4.5 dB	TO-72	3/R	VHF/UHF/RF amp.mix., osc.
3N202	N-Dual-Gate MOSFET	360	25	50	8K	6	-5	200	200 MHz 4.5 dB	TO-72	3/S	VHF amp., mixer
MPF102	N-JFET	310	25	20	2K	4.5	-8	200		TO-92	2/N,M	HF/VHF amp.,mix., osc.,
MPF106/ 2N5484	N-JFET	310	25	30	2.5K	5	-6	400	200 MHz 4 dB	TO-92	2/N,M	HF/VHF/UHF amp.,mix.,osc.
40673	N-Dual-Gate MOSFET	330	20	50	12K	6	-4	400	200 MHz 6 dB	TO-72	3/R	HF/VHF/UHF amp. mix., osc.
U300	P-JFET	300	-40	20	8K	-50	+10	-	400 MHz	TO-18	4/S	General Purpose amp.
U304	P-JFET	350	-30	-50	10K	27	+10	-	-	TO-18	4/S	analog switch, chopper
U310	N-JFET	500	30	60	10K	2.5	-6	450	450 MHz 3.2 dB	TO-52	5/S	common-gate VHF/UHF amp.,osc., mixer
U350	300 N-JFET Quad	30 1W	25	60	9K	5	-6	100	100 MHz 7 dB	TO-99	6/S	matched JFET doubly bal. mixer
U431	N-JFET Dual	300	25	30	10K	5	-6	100	-	TO-99	7/S	matched JFET cascade amp., balanced mixer

* 25°C   S = Siliconix Inc.   R = RCA   N = National Semiconductor   M = Motorola

Three-Terminal Voltage Regulators

* Listed numerically by device

Device	Description	Voltage	Current (Amps)	Package
317	Adj. Pos	+1.2 to +37	0.5	TO-205
317	Adj. Pos	+1.2 to +37	1.5	TO-204, TO-220
317L	Low Current Adj. Pos	+1.2 to +37	0.1	TO-205, TO-92
317M	Med Current Adj. Pos	+1.2 to +37	0.5	TO-220
350	High Current Adj. Pos	+1.2 to +33	3.0	TO-204, TO-220
337	Adj. Neg	-1.2 to -37	0.5	TO-205
337	Adj. Neg	-1.2 to -37	1.5	TO-204, TO-220
337M	Med Current Adj. Neg	-1.2 to -37	0.5	TO-220
309		+5	0.2	TO-205
309		+5	1.0	TO-204
323		+5	9.0	TO-204, TO-220
140-XX	Fixed Pos	**Note #**	1.0	TO-204, TO-220
340-XX			1.0	TO-204, TO-220
78XX			1.0	TO-204, TO-220
78LXX			0.1	TO-205, TO-92
78MXX			0.5	TO-220
78TXX			3.0	TO-204
79XX	Fixed Neg	**Note #**	1.0	TO-204, TO-220
79LXX			0.1	TO-205, TO-92
79MXX			0.5	TO-220

Legend:

Adj.	= Adjustable
Med	= Medium
Neg	= Negative
Pos	= Positive

Note # - XX indicates the regulated voltage; which may be anywhere from 1.2 volts to 35 volts. For example a 7808 is a positive 8-volt regulator, and a 7912 is a negative 12-volt regulator.

The regulator package may be denoted by an additional suffix, according to the following:

Package	Suffix
TO-204 (TO-3)	K
TO-220	T
TO-205 (TO-39)	H,G
TO-92	P,Z

Example:

A 7815K is a positive 15-volt regulator in a TO-204 package. An LM340T-8 is a positive 8-volt regulator in a TO-220 package. In addition, different manufacturers use different prefixes. An LM7812 is equivalent to a µA 7812 or MC7812.

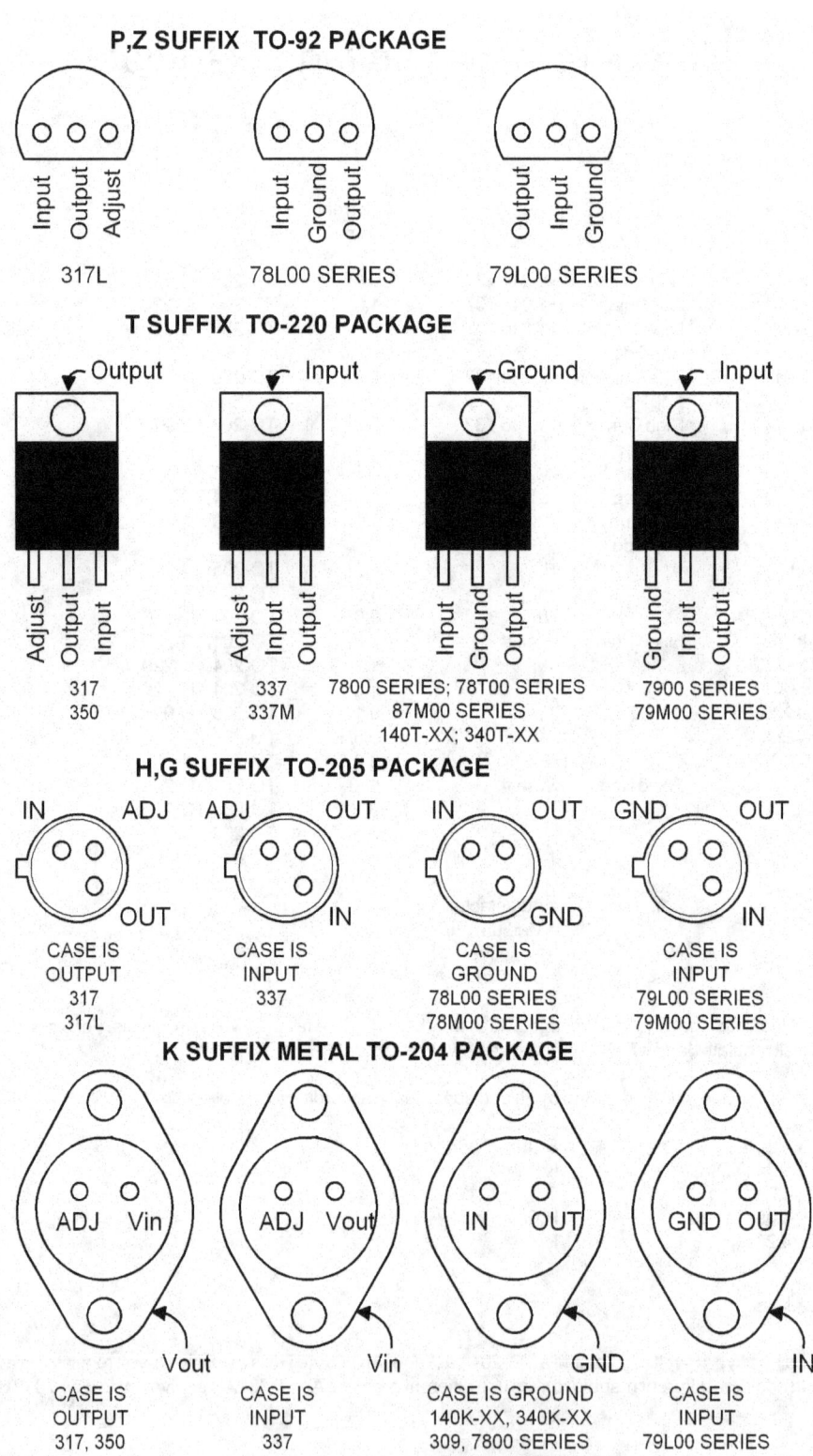

P,Z SUFFIX TO-92 PACKAGE

Input Output Adjust	Input Ground Output	Output Input Ground
317L	78L00 SERIES	79L00 SERIES

T SUFFIX TO-220 PACKAGE

Output	Input	Ground	Input
Adjust Output Input	Adjust Input Output	Input Ground Output	Ground Input Output
317 350	337 337M	7800 SERIES; 78T00 SERIES 87M00 SERIES 140T-XX; 340T-XX	7900 SERIES 79M00 SERIES

H,G SUFFIX TO-205 PACKAGE

IN ADJ / OUT	ADJ OUT / IN	IN OUT / GND	GND OUT / IN
CASE IS OUTPUT 317 317L	CASE IS INPUT 337	CASE IS GROUND 78L00 SERIES 78M00 SERIES	CASE IS INPUT 79L00 SERIES 79M00 SERIES

K SUFFIX METAL TO-204 PACKAGE

ADJ Vin / Vout	ADJ Vout / Vin	IN OUT / GND	GND OUT / IN
CASE IS OUTPUT 317, 350	CASE IS INPUT 337	CASE IS GROUND 140K-XX, 340K-XX 309, 7800 SERIES 78T00 SERIES	CASE IS INPUT 79L00 SERIES

Suggested Books for Further Reading

Do-It-Yourself Circuitbuilding For Dummies (Do-It-Yourself for Dummies) (Paperback) by H. Ward Silver, Published by For Dummies (2008)

MORE Electronic Gadgets for the Evil Genius (Paperback) by Rober Ianninni, Published by McGraw-Hill/TAB Electronics; 2 edition (December 20, 2005)

The Art of Electronics by Paul Horowitz and Winfried Hill, Published by Cambridge University Press; 2 edition (July 28, 1989)

Electronics For Dummies (Paperback) by Gordon McComb and Earl Boysen, Published by For Dummies (2005)

Practical Electronics for Inventors (Paperback) by Paul Scherz, Published by McGraw-Hill/ TAB Electronics; 2 edition (September 1, 2006)

Teach Yourself Electricity and Electronics, Fourth Edition (Teach Yourself) (Paperback) by Stan Gibilisco, McGraw-Hill/TAB Electronics; 4 edition (March 15, 2006)

The Circuit Designer's Companion, Second Edition (Paperback) by Tim Williams, Published by Newnes; 2 edition (December 21, 2004)

Circuit-Bending: Build Your Own Alien Instruments (ExtremeTech) (Paperback) by Reed Ghazala, Published by Wiley (August 26, 2005)

The Tube Amp Book: Deluxe Revised Edition (Spiral-bound) by Aspen Pittman, Published by Backbeat Books; Har/Cdr edition (September 1, 2003)

Circuit Design with VHDL (Hardcover) by Volnei A. Pedroni, Published by The MIT Press (August 1, 2004)

Active Filter Cookbook, Second Edition (Paperback) by DON LANCASTER. Newnes; 2 edition (August 13, 1996)

123 PIC Microcontroller Experiments for the Evil Genius (Paperback) by Myke Predko. McGraw-Hill/TAB Electronics; 1 edition (June 24, 2005)

Bionics for the Evil Genius (Paperback) by Newton C. Braga. McGraw-Hill/TAB Electronics; 1 edition (December 22, 2005)

Electronic Circuits for the Evil Genius (Paperback) by Dave Cutcher . McGraw-Hill/TAB Electronics; 1 edition (November 24, 2004)

Semiconductor Optoelectronic Devices (2nd Edition) (Paperback) by Pallab Bhattacharya Prentice Hall; 2 edition (November 29, 1996)

Suggested Books for Further Reading

Cooling Techniques for Electronic Equipment, 2nd Edition (Hardcover) by Dave S. Steinberg. Wiley-Interscience; 2 edition (October 8, 1991)

Mastering Audio, Second Edition: The art and the science (Paperback) by Bob Katz. Focal Press; 2 edition (October 1, 2007)

101 Spy Gadgets for the Evil Genius (Paperback) by Brad Graham & Kathy McGowan. McGraw-Hill/TAB Electronics; 1 edition (June 19, 2006)

Solar Energy Projects for the Evil Genius (Paperback) by Gavin D J Harper. McGraw-Hill/ TAB Electronics; 1 edition (June 22, 2007)

Tab Electronics Guide to Understanding Electricity and Electronics (Paperback) by G. Randy Slone. McGraw-Hill/TAB Electronics; 2nd edition (July 21, 2000)

Electronics Sensors for the Evil Genius: 54 Electrifying Projects (Evil Genius) (Paperback) by Thomas Petruzzellis. McGraw-Hill/TAB Electronics; 1 edition (January 2006)

Beginner's Guide to Reading Schematics (Paperback) by Robert J. Traister & Anna L. Lisk. McGraw-Hill/TAB Electronics; 2 edition (March 1, 1991)

How to Test Almost Everything Electronic (Paperback) by Delton T. Horn. McGraw-Hill/TAB Electronics; 1 edition (April 1, 1993)

Crystal Fire: The Invention of the Transistor and the Birth of the Information Age by Michael Riordan & Lillian Hoddeson. W. W. Norton & Company; New Ed edition (1998)

Electronic Gadgets for the Evil Genius : 28 Build-It-Yourself (Paperback) by Robert E. Iannini. McGraw-Hill/TAB Electronics; 1 edition (March 12, 2004)

High-Power Audio Amplifier Construction Manual (Paperback) by G. Randy Slone. McGraw-Hill/TAB Electronics; 1 edition (May 1, 1999)

Mechatronics for the Evil Genius (Paperback) by Newton C. Braga. McGraw-Hill/TAB Electronics; 1 edition (September 15, 2005)

Industrial Motor Control (Hardcover) by Stephen L. Herman. CENGAGE Delmar Learning; 4 edition (October 8, 1998)

Audio Electronics, Second Edition (Paperback) by John Linsley Hood. Newnes; 2 edition (November 16, 1998)

PRINTED CIRCUIT BOARD LAYOUTS

All printed circuit board layouts in this collection are once again printed in the following pages. You can either cut out or photocopy these pages to make a separate file for quick reference.

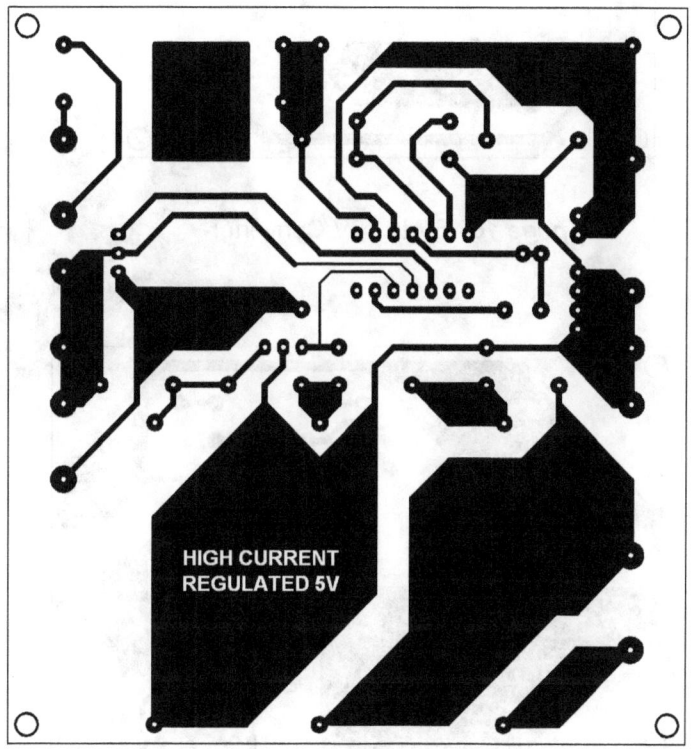

page 11   High Current Regulated 5V

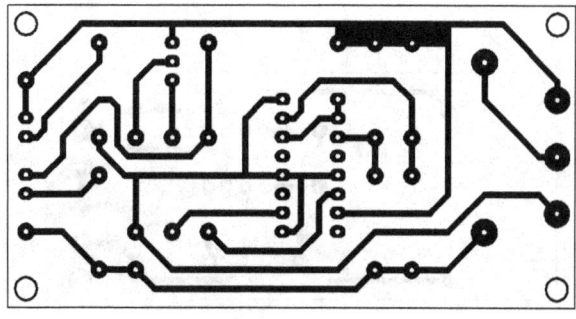

page 14   Voltage Monitor

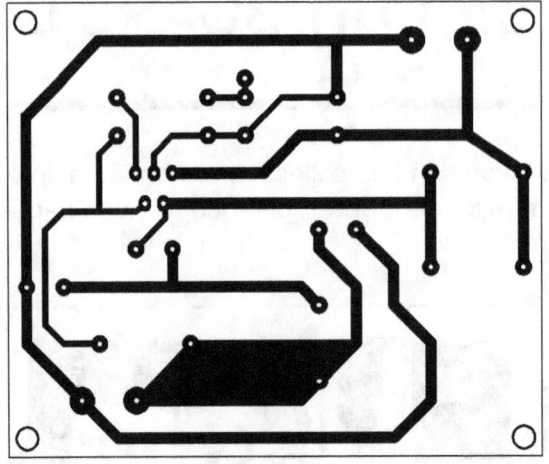

page 16 6V to 12V Converter

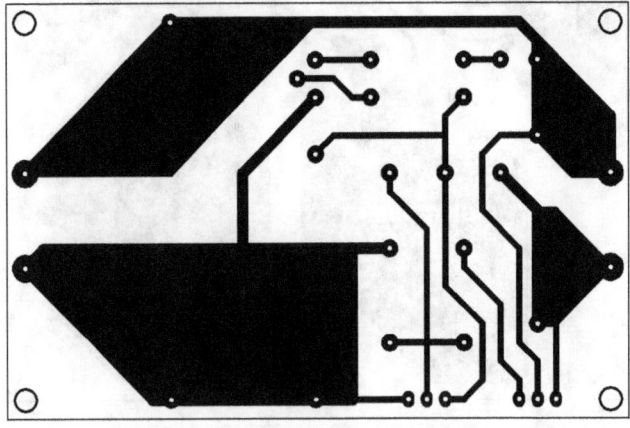

page 18 5V/3A Power Source

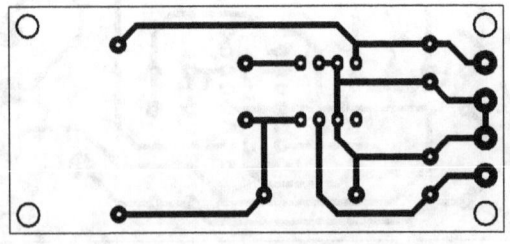

page 26 Simple Switching Supply

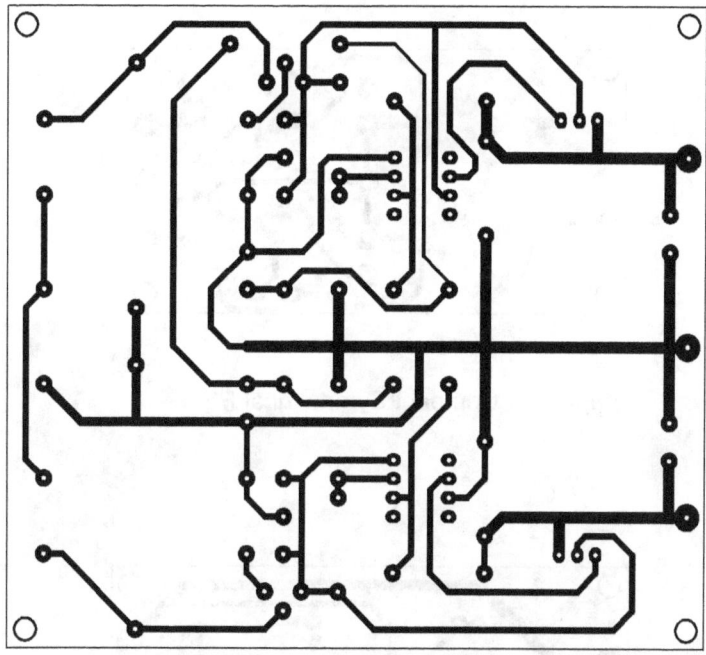

page 20   Symmetrical Power Supply

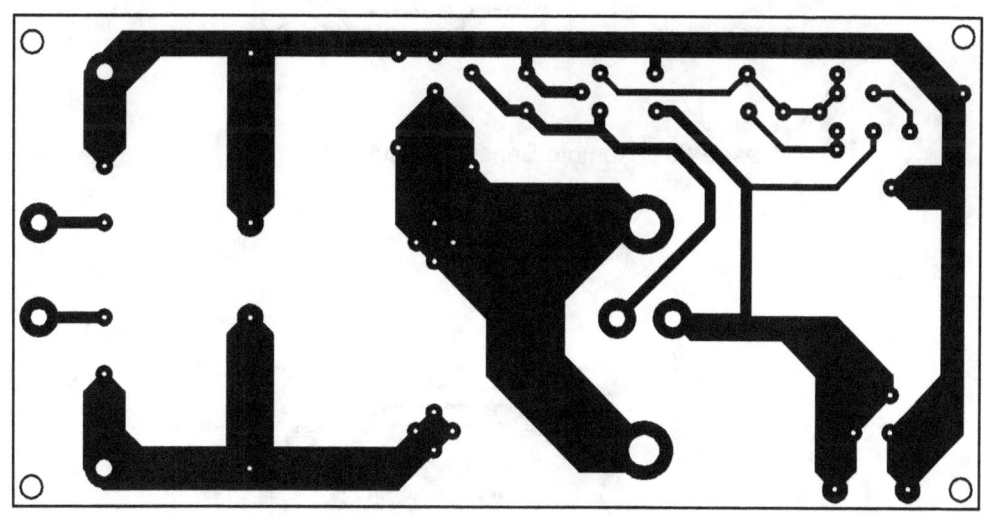

page 22   Regulated Power Supply

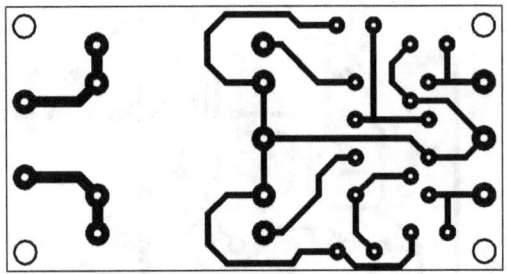

page 27   Compact Symmetrical PS

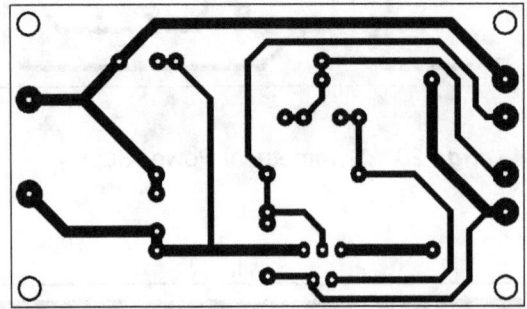

page 29   Remote Sensing Regulator

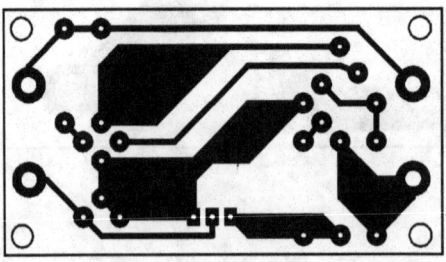

page 35   Power Supply Regulator

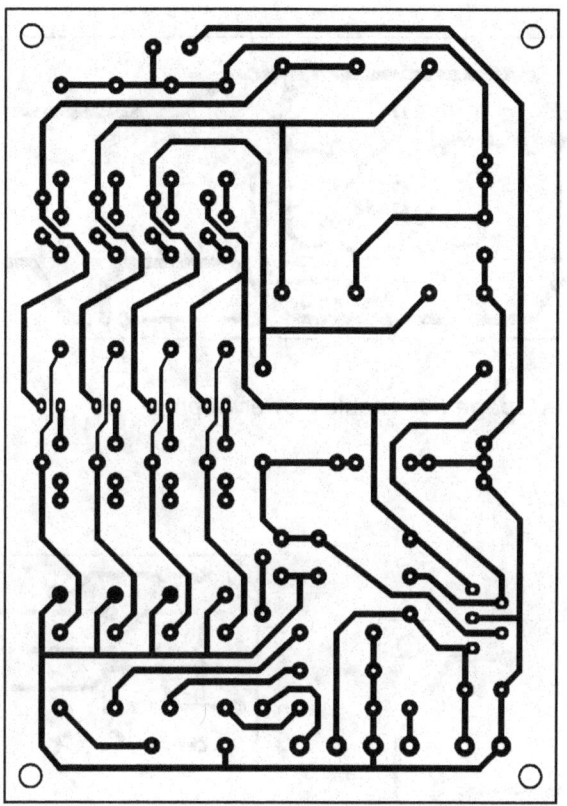

page 31   Current Monitored Supply

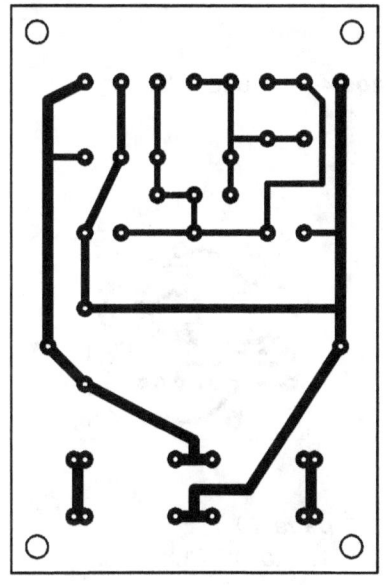

page 37   Variable Power Supply

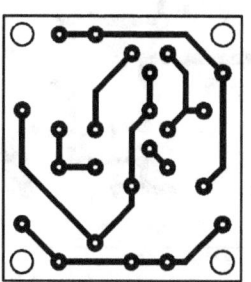

page 41   Stable Zener Voltage

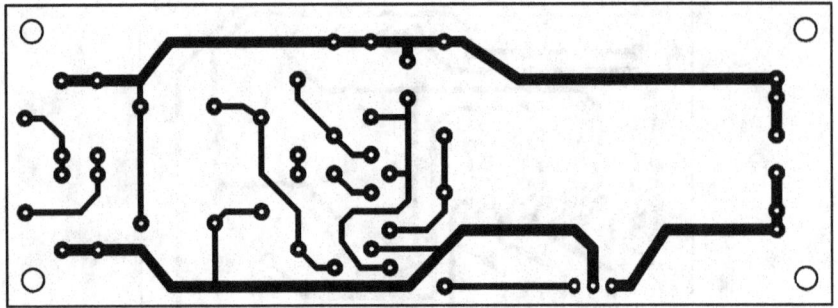

page 39   Stable Power Supply

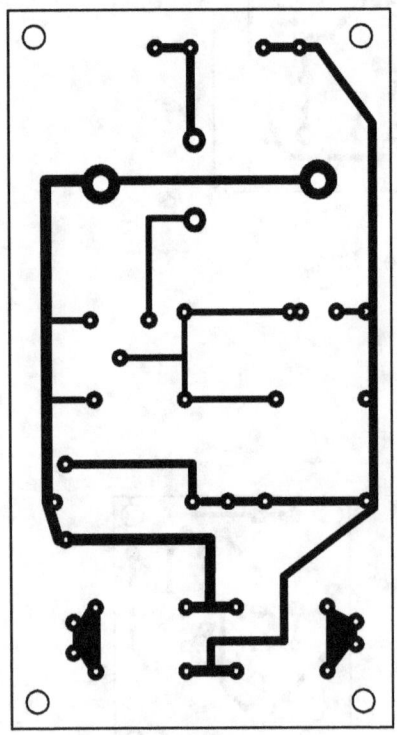

page 42   3-Watt Amp Power Supply

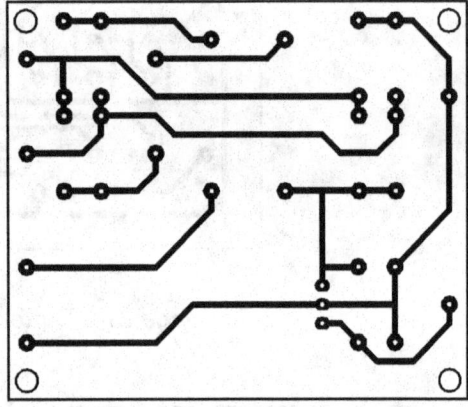

page 44   Standby Supply

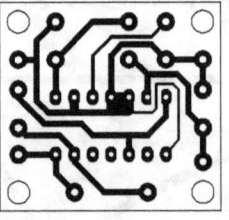

page 49
DC to DC Converter

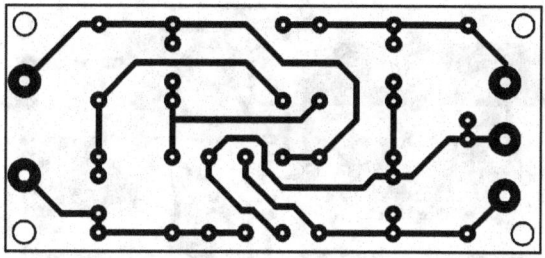

page 46   Symmetrical Aux PS

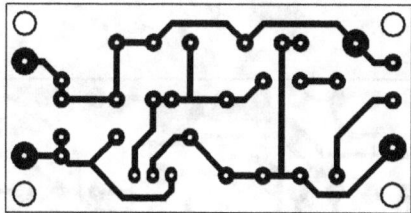

page 50   3-Ampere Power Supply

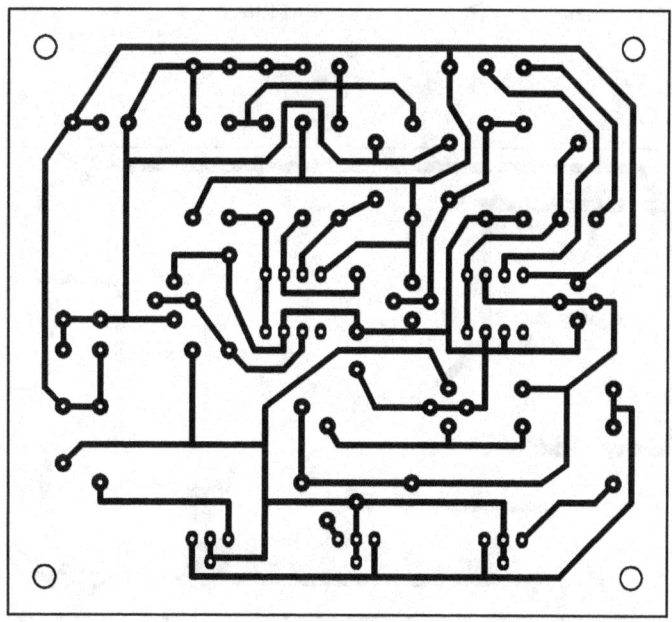

page 52   0...50V/0...2A Power Supply

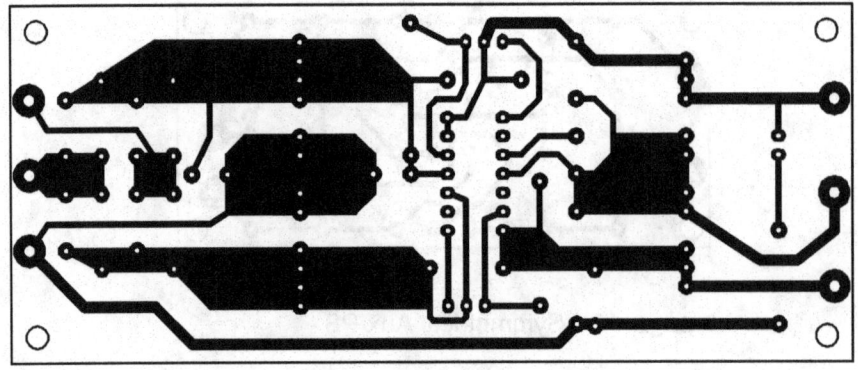

page 55 Supply for Opamps

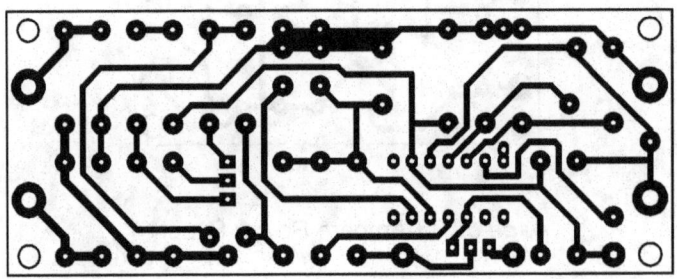

page 57 PS w/ Dissipation Limiter

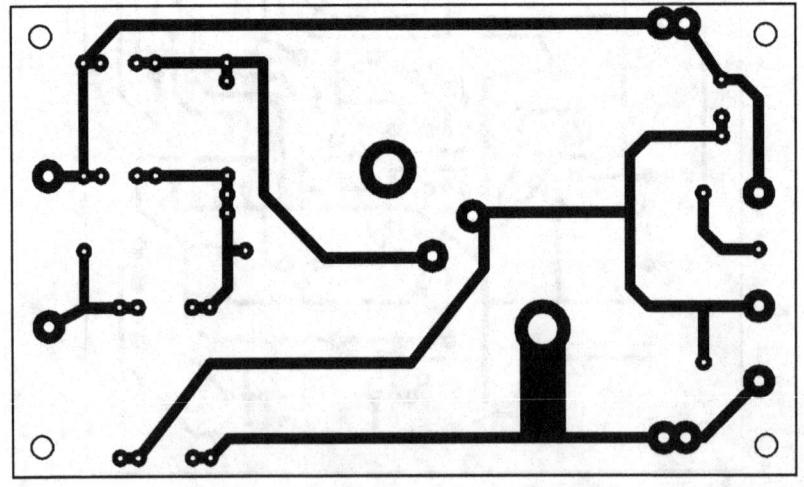

page 59 5-A Delayed Power On

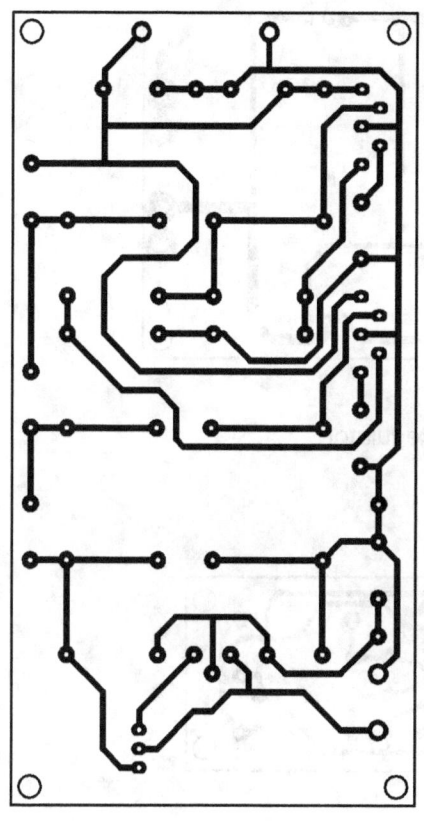

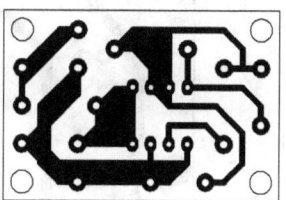

page 65
Versatile Power Supply

page 62   6V to 12V Converter

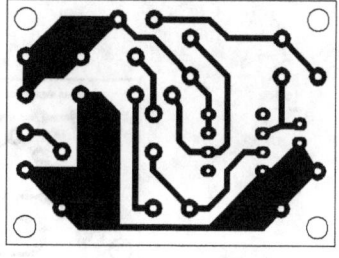

page 71   Low Drop Regulator

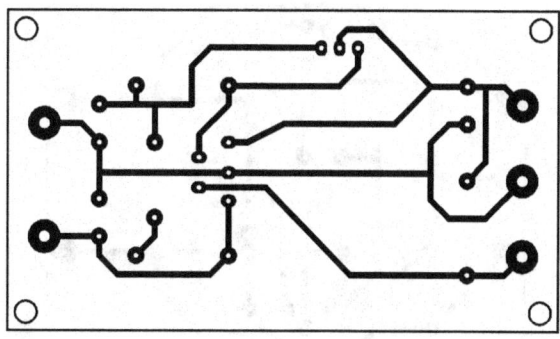

page 67   Robust 5V Supply

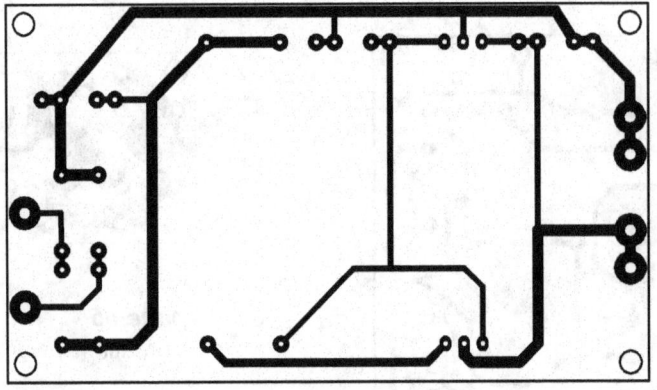

page 73   Amplified Regulator

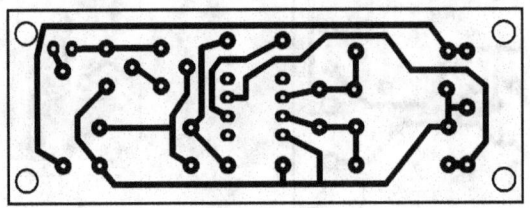

page 76   78XX Regulator Monitor

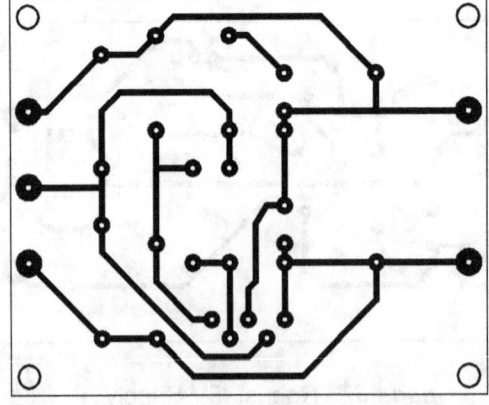

page 85   Symmetrical Power Supply

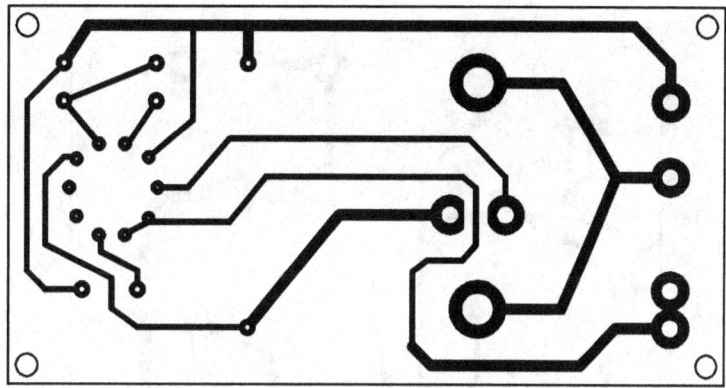

page 81   Constant Current

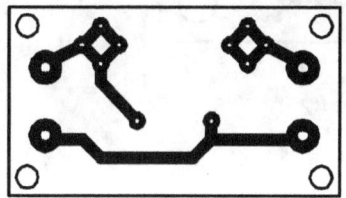

page 79   Overvoltage Crowbar #1

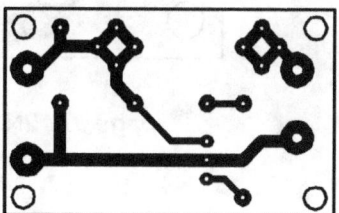

page 79   Overvoltage Crowbar #2

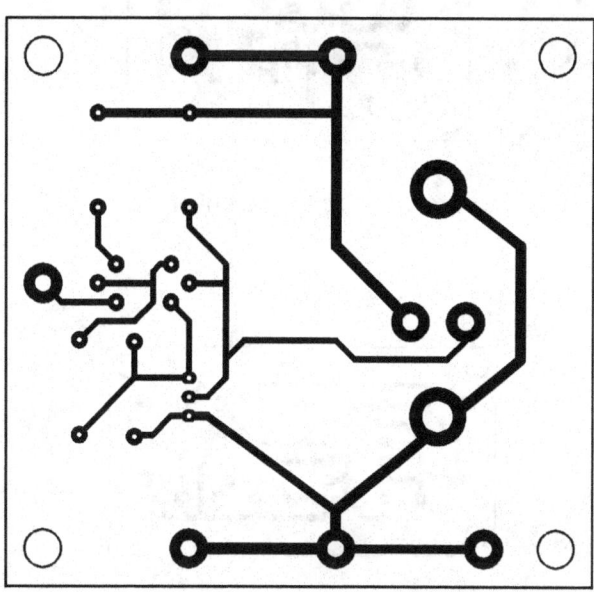

page 87   2N3055 Darlingtons #1

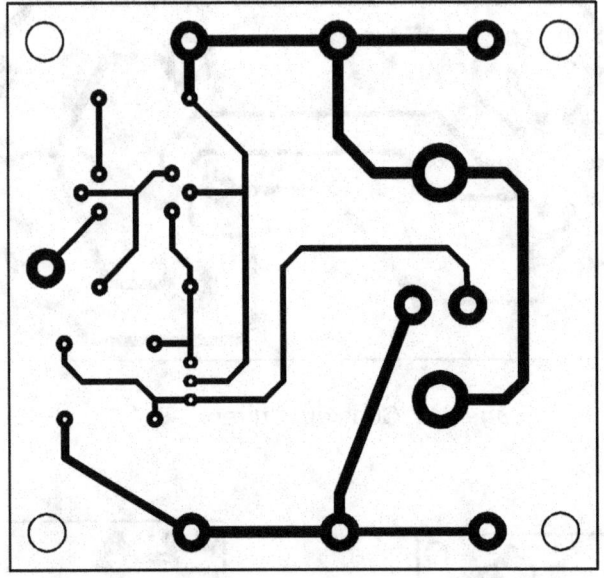

page 87 2N3055 Darlingtons #2

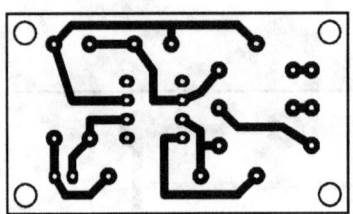

page 90 Voltage Monitor

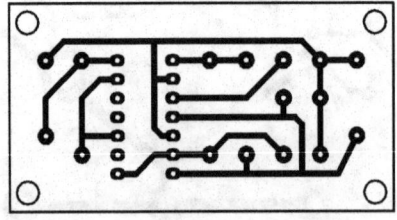

page 93 Polarity Inverter

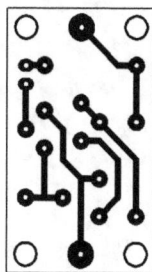

page 91   Temperature Sensor

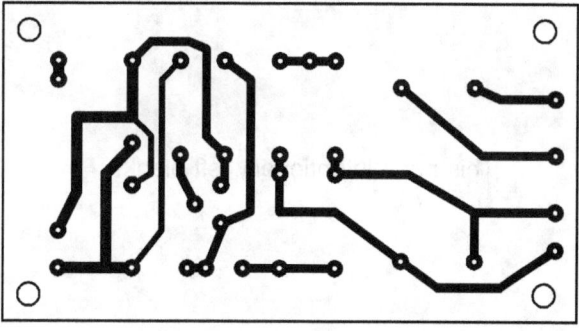

page 99   Current Alarm #1

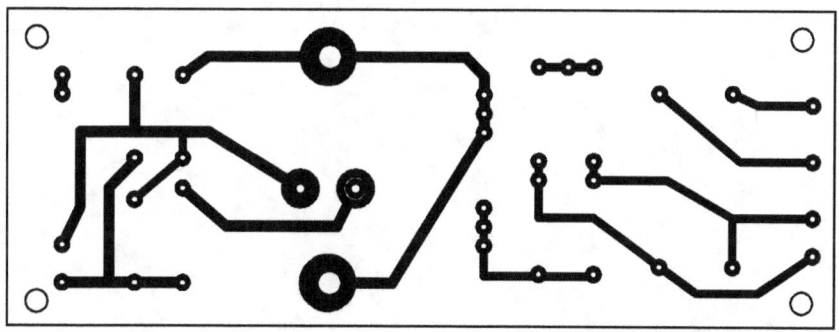

page 99   Current Alarm #2

This page intentionally left blank.

Index

Index

Index

Notes

Notes

Notes

Notes

Notes

ONE HUNDRED electronic circuits with
ready to use pcb design layouts!

Get your copy now from
amazon.com!

ONE HUNDRED and TWO circuits with
ready to use pcb design layouts!

Get your copy now from
amazon.com!

This authoritative and well-researched book is the only one available that will give you all of the most important and reliable on VHF antenna construction techniques.

This unique book offers a superb collection of detailed, easy-to-follow, fully illustrated, and tested designs, covering such types of antennas as:

Omnidirectional antennas

Gain-omni antennas

Gain-directed beams

Portable antennas

Yagi antennas

Stacked arrays

Stacked collinears

Wideband-omni antennas

Get your copy now from amazon.com!

A compilation of audio circuits with
ready to use pcb design layouts!

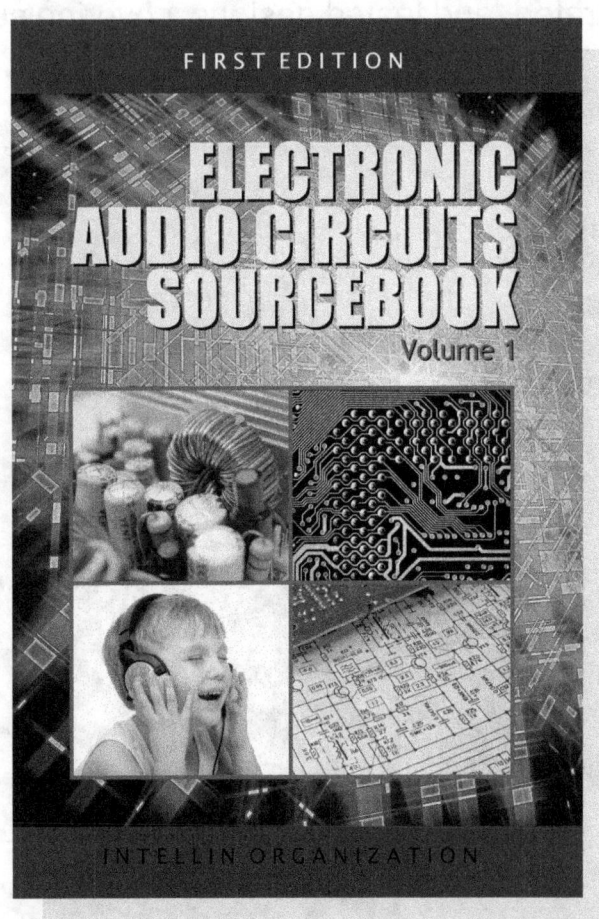

Get your copy now from
amazon.com!

www.ingramcontent.com/pod-product-compliance
Lightning Source LLC
Chambersburg PA
CBHW081128170526
45165CB00008B/2590